Cosmos

The infographic book of space

First published 2015 by Aurum Press Ltd
74-77 White Lion Street, London N1 9PF

This paperback edition first published in 2017

www.quartoknows.com

A catalogue record for this book is
available from the British Library.

ISBN 978 1 78131 645 0

Printed in China.

Design by **Founded**.
www.wearefounded.com

FSC
www.fsc.org
MIX
Paper from
responsible sources
FSC® C008047

Contents

Introduction

Space and astronomy are subjects that can really grab the imagination - something they did to both of us at an early age. Many of the detailed explanations may seem complex, and sometimes rather subtle, but the basic ideas are familiar to us all at some level. The scales and distances can be so vast as to be all but unimaginable, and simply writing out huge numbers is not always helpful.

In this book we've tried to display the processes and concepts in a visual way, allowing the ideas to be easily viewed without hiding the details. Where possible we show the data to scale. For example, in 'Journey to the Moon' the Earth, Moon and the size of the Moon's orbit are all to scale. However, given the vast range of sizes and ideas in astronomy, that isn't always achievable within the confines of the page. So there are instances where we have used logarithmic scaling or, in the most extreme cases, abstracted the scale entirely.

We've ranged from human exploration of the Earth and Moon, to the way galaxies are scattered throughout the cosmos over scales of billions of light years; from the building of telescopes to observe the heavens, to

humanity's attempts to contact alien civilisations. Whatever your knowledge of space and astronomy, there's bound to be something to interest you.

The graphics are based on as up-to-date knowledge and research as possible. Most of the data sets are valid up to the end of 2014. As is the nature of a field of active research, new discoveries are made and our knowledge moves on so some aspects could very well be out-of-date by the time of printing. We'll keep track of updates, as well as provide interactive versions of some infographics, at cosmos-book.github.io.

While both of us are astronomers, our professional research has been focussed on relatively narrow fields, so some areas of the book were relatively new to us when we started. We have, however, both hugely enjoyed public communication of astronomy, from podcasts and websites, to radio and TV programmes. This outreach work spans fields spread across almost all of astronomy, but despite this experience we have both learned an awful lot compiling this book. We hope you enjoy reading it as much as we enjoyed writing it!

Stuart Lowe & Chris North
March 2015

One / Space exploration

Launch vehicles

If you want to launch something into space you have a range of options,
from government space agencies to private companies. The cost depends
on how much you want to send up, how far up you want to send it, and
how much risk you are willing to take.

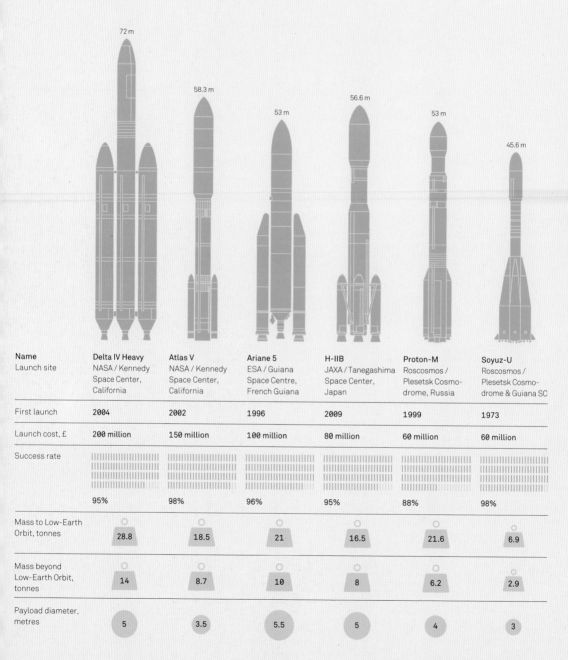

Name Launch site	Delta IV Heavy NASA / Kennedy Space Center, California	Atlas V NASA / Kennedy Space Center, California	Ariane 5 ESA / Guiana Space Centre, French Guiana	H-IIB JAXA / Tanegashima Space Center, Japan	Proton-M Roscosmos / Plesetsk Cosmo- drome, Russia	Soyuz-U Roscosmos / Plesetsk Cosmo- drome & Guiana SC
First launch	2004	2002	1996	2009	1999	1973
Launch cost, £	200 million	150 million	100 million	80 million	60 million	60 million
Success rate	95%	98%	96%	95%	88%	98%
Mass to Low-Earth Orbit, tonnes	28.8	18.5	21	16.5	21.6	6.9
Mass beyond Low-Earth Orbit, tonnes	14	8.7	10	8	6.2	2.9
Payload diameter, metres	5	3.5	5.5	5	4	3

	63.4 m	38.2 m	54.8 m	30 m	16.9 m	44 m

Name	Falcon 9	Delta II	Long March 3B	Vega	Pegasus	PSLV
Launch site	NASA / Kennedy Space Center, California	NASA / Kennedy Space Center, California	CNSA / Xichang Satellite Launch Center, China	ESA / Guiana Space Centre, French Guiana	Orbital / Carrier Aircraft	ISRO / Satish Dhawan Space Centre, India
First launch	2013	1990	1996	2012	1990	1993
Launch cost, £	40 million	30 million	30 million	23 million	15 million	11.4 million
Success rate	97%	99%	75%	98%	92%	96%
Mass to Low-Earth Orbit, tonnes	13.2	5	11.5	1.4	0.4	3.8
Mass beyond Low-Earth Orbit, tonnes	4.9	1.8	5.5	0	0	1.3
Payload diameter, metres	3.5	3	3.5	3	1.2	3.2

One small step for...

Humans are not the only, or even the first, species to venture into space. The first recorded space flight was in 1947. Those pioneering astronauts were fruit flies and they returned alive. In 1949 they were followed by the first monkeys in space although it wasn't until 1959 that 'Able' and 'Baker' became the first monkeys to survive spaceflight. In 1951 mice were the first mammals to survive actual space-flight conditions. Close behind, dogs made their first successful trips to space in 1951

and made their first orbital flight in 1957. In March 1961, mice (together with frogs, guinea pigs, and insects) beat humans by a few weeks to be the first animals to successfully orbit the Earth.

In September 1968, three months before *Apollo 8*, *Zond 5* took the first earthlings around the Moon and returned them safely to Earth. The crew included a tortoise, wine flies and meal worms.

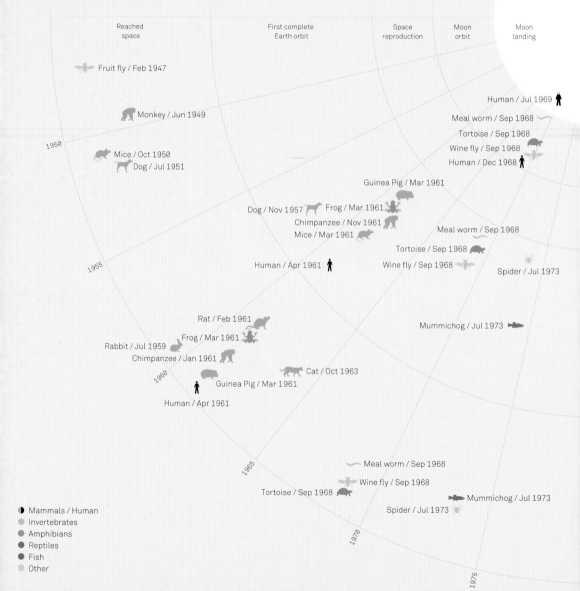

Reached space · First complete Earth orbit · Space reproduction · Moon orbit · Moon landing

Fruit fly / Feb 1947
Monkey / Jun 1949
1950
Mice / Oct 1950
Dog / Jul 1951
1955

Human / Jul 1969
Meal worm / Sep 1968
Tortoise / Sep 1968
Wine fly / Sep 1968
Human / Dec 1968

Guinea Pig / Mar 1961
Dog / Nov 1957 Frog / Mar 1961
Chimpanzee / Nov 1961
Mice / Mar 1961
Meal worm / Sep 1968
Tortoise / Sep 1968
Human / Apr 1961 Wine fly / Sep 1968
Spider / Jul 1973

Rat / Feb 1961
Frog / Mar 1961
Rabbit / Jul 1959
Chimpanzee / Jan 1961
1960
Cat / Oct 1963
Human / Apr 1961 Guinea Pig / Mar 1961
Mummichog / Jul 1973

1965
Meal worm / Sep 1968
Wine fly / Sep 1968
Tortoise / Sep 1968
Mummichog / Jul 1973
Spider / Jul 1973

1970

1975

- Mammals / Human
- Invertebrates
- Amphibians
- Reptiles
- Fish
- Other

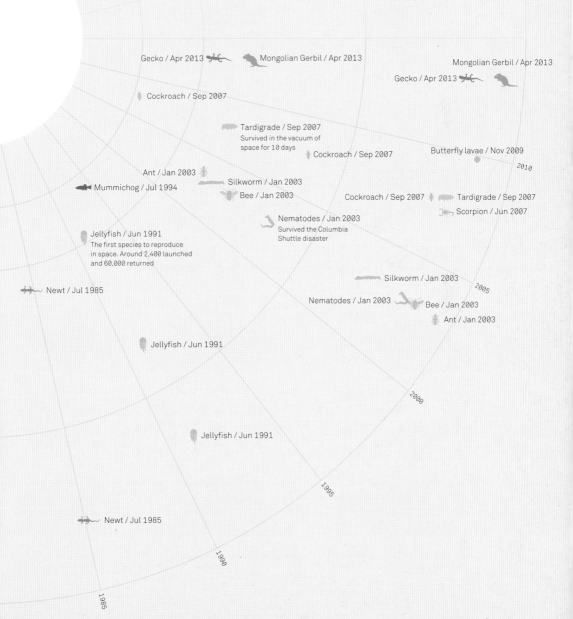

Gecko / Apr 2013

Mongolian Gerbil / Apr 2013

Mongolian Gerbil / Apr 2013

Gecko / Apr 2013

Cockroach / Sep 2007

Tardigrade / Sep 2007
Survived in the vacuum of
space for 10 days

Cockroach / Sep 2007

Butterfly lavae / Nov 2009

2010

Ant / Jan 2003

Silkworm / Jan 2003

Mummichog / Jul 1994

Bee / Jan 2003

Cockroach / Sep 2007

Tardigrade / Sep 2007

Scorpion / Jun 2007

Nematodes / Jan 2003
Survived the Columbia
Shuttle disaster

Jellyfish / Jun 1991
The first species to reproduce
in space. Around 2,400 launched
and 60,000 returned

Silkworm / Jan 2003

2005

Newt / Jul 1985

Nematodes / Jan 2003

Bee / Jan 2003

Ant / Jan 2003

Jellyfish / Jun 1991

2000

Jellyfish / Jun 1991

1995

Newt / Jul 1985

1990

1985

Human spaceflight

Humans first reached space (defined as 100 kilometres above the surface) in 1961 with the flight of cosmonaut Yuri Gagarin (USSR). The first woman – Valentina Tereshkova (USSR) – followed in 1963. Far from peaking at the time of Apollo, the number of people in space gradually grew during the 1980s and 1990s with the Mir Space Station and Shuttle programmes. Since 31 October 2000 humanity has had a continuous presence in space in the form of the permanently staffed International Space Station.

Thankfully, considering the dangers involved, there have been relatively few deaths in space. In 1967 Vladimir Komarov (USSR) died on impact after a parachute failure during re-entry. Georgi Dobrovolski (USSR), Viktor Patsayev (USSR) and Vladislav Volkov (USSR) all died in 1971 after they undocked from the *Salyut 1* space station to return to Earth. The Space Shuttle Challenger exploded on launch in 1986 killing Greg Jarvis (US), Christa McAuliffe (US), Ronald McNair (US), Ellison Onizuka (US), Judith Resnik (US), Michael Smith (US) and Dick Scobee (US). In 2003 the Space Shuttle Columbia broke up on re-entry, due to a damaged heat tile, killing Michael Anderson (US), David Brown (US), Kalpana Chawla (US), Laurel Clark (US), Rick Husband (US), William McCool (US), and Ilan Ramon (Israel). Both Shuttle accidents resulted in pauses on human spaceflight as the accidents were investigated.

♀ Female ♂ Male ♦ Fatality
Name, country | notable firsts by country

	Year
	1961
	1962
Valentina Tereshkova, USSR 1 ♀	1963
	1964
	1965
	1966
	1967
	1968
US Apollo 11 /	1969
	1970
	1971
	1972
Skylab /Start	1973
	1974
	1975
	1976
	1977
	1978
	1979
Skylab / End	1980
	1981
1 ♂	1982
1 ♂	1983
Judith Resnik, US 5 ♀	♂ ♂ ♂ ♂	1984
3 ♂ ♂ ♂	1985
Mir Space Station / Start	
Challenger Disaster /	1986
	1987
	1988
4 ♂ ♂ ♂ ♂	1989
3 ♂ ♂ ♂	1990
Helen Sharman, UK 6 ♀	♂ ♂ ♂ ♂ ♂	1991
8 ♂ ♂ ♂ ♂ ♂ ♂ ♂ ♂	1992
7 ♂ ♂ ♂ ♂ ♂ ♂ ♂	1993
Chiaki Mukai, Japan 8 ♀	♂ ♂ ♂ ♂ ♂ ♂ ♂	1994
10 ♂ ♂ ♂ ♂ ♂ ♂ ♂ ♂ ♂ ♂	1995
Claudie Haigneré, France 5 ♀	♂ ♂ ♂ ♂	1996
10 ♂ ♂ ♂ ♂ ♂ ♂ ♂ ♂ ♂ ♂	1997
International Space Station / Start	
6 ♂ ♂ ♂ ♂ ♂ ♂	1998
5 ♂ ♂ ♂ ♂ ♂	1999
4 ♂ ♂ ♂ ♂	2000
5 ♂ ♂ ♂ ♂ ♂	2001
Mir Space Station / End	
4 ♂ ♂ ♂ ♂	2002
Columbia Disaster / 3 ♦ ♦ ♦	2003
	2004
2 ♂ ♂	2005
7 ♂ ♂ ♂ ♂ ♂ ♂ ♂	2006
5 ♂ ♂ ♂ ♂ ♂	2007
5 ♂ ♂ ♂ ♂ ♂	2008
3 ♂ ♂ ♂	2009
4 ♂ ♂ ♂ ♂	2010
2 ♂ ♂	2011
Yang Liu, China 1 ♀	2012
1 ♂	2013
2 ♂ ♂	2014

- 4 Yuri Gagarin, USSR & Alan Shepard, US
- 5
- 2
- 3
- 11
- 9
- 1
- 7
- 23 Neil Armstrong, US
- 5
- 12
- 6
- 16
- 6
- 8
- 6
- 8
- 10 Sigmund Jähn, Germany
- 4
- 13
- 10
- 15 Jean-Loup Chrétien, France
- 24
- 31
- 53
- 9
- 10
- 22
- 25
- 35 Toyohiro Akiyama, Japan
- 34
- 51 Franco Malerba, Italy
- 40
- 43
- 40
- 43
- 51
- 33
- 15
- 33
- 41
- 35
- 11 Liwei Yang, China
- 6
- 14
- 22
- 21
- 36
- 42
- 26
- 27
- 15
- 16
- 9

Travels in time and space

In the 20th Century space travel left the realms of science fiction. These days it is almost routine, even if currently confined to a Low-Earth Orbit. Astronauts regularly spend many months at a time in orbit. At orbital velocity, they circle our planet 16 times a day and cover huge distances.

One interesting feature of travelling at orbital speeds is that time runs slightly slower than it does for those of us on the ground. The result is that astronauts are ever so slightly younger than they'd be if they stayed at home. This effect is tiny (25 milliseconds at most) but is comparable to the gap between the world's top six fastest 100-metre runners.

A   **Neil Alden Armstrong, US** / First flight 1966 /
Days in space 8.58 / First person to walk on the Moon.

B   **Edward Michael Fincke, US** / First flight 2004 /
Days in space 381.63 / American with most time
in space – 381.63 days.

C   **Yuri Alekseyevich Gagarin, USSR** / First flight 1961 /
Days in space 0.08 / The first person in space – 1961.

D   **Sergei Konstantinovich Krikalyov, USSR** / First flight 1988 /
Days in space 803.4 / Has spent the most time in space – 803.4.

E   **Valeri Vladimirovich Polyakov, USSR** / First flight 1988 /
Days in space 678.69 / Longest single flight at 437.75 days.

F   **Charles Simonyi, Hungary** / First flight 2007 /
Days in space 26.6 / The space tourist with the most time in space.

G   **Anatoli Yakovlevich Soloviyov, USSR** / First flight 1988 /
Days in space 651 / Has spent the most time on space
walks – 68 hours 44 minutes.

H   **Dennis Tito, US** / First flight 2001 /
Days in space 7.92 / First space tourist.

I   **Koichi Wakata, Japan** / First flight 1996 /
Days in space 238.24 / International astronaut with the most
time in space – 238.24 days.

J   **Peggy Annette Whitson, US** / First flight 2002 /
Days in space 376.72 / Woman with the most time
in space – 376.72 days.

K   **Liwei Yang, China** / First flight 2003 /
Days in space 0.89 / First Chinese taikonaut.

- US Astronauts
- Russian Cosmonauts
- Chinese Taikonauts
- International Astronauts
- Space Tourists

C

K

0.1 days in space   1 day

10 days   100 days   1,000

Space survival

In TV and movies, people who find themselves suddenly exposed to the vacuum of space often explode or instantly freeze to death. Neither of these scenarios will occur and death won't be instantaneous. We have some idea what will happen based on tests of animals (including humans) as well as accidents that occurred in pressure chambers on the ground and during space-flight.

You will not freeze to death / You will not immediately freeze. Space is a fairly good insulator so there is no conduction/convection. In sunlight, in orbit around Earth, you may radiate energy slightly quicker than you would at room temperature. You will cool slowly.

Blood will not boil / Unless you go into deep shock, the pressure will stay high enough that it won't boil.

Sunburn / If you have no protection, UV radiation from the Sun will give you very bad sunburn.

Partial exposure / If only part of your body is exposed, your chances are better. In 1960 Joe Kittinger Jr exposed his right hand to low pressure on a high-altitude balloon flight. It swelled to twice its size but returned to normal within a few hours.

Sound / After the initial loss of air you won't be able to hear anything.

Abdominal distress / Expansion of gases in your stomach may cause pain. The advice is to pass any excess gas.

A punctured spacecraft / If your spacecraft is ten cubic metres in volume and gets a one square centimetre puncture, it'll take about six minutes for the pressure to drop to half an atmosphere where critical hypoxia starts.

From the moment of the explosive decompression you'll have about 60 – 90 seconds to get back into a pressurised atmosphere if you are to survive. The clock is ticking...

B
You'll experience an initial rise in heart rate due to the shock. Any adrenaline rush will use up your oxygen more quickly. Stay calm. Obviously that is much easier said than done.

A
The very first thing to do is to breathe out. If you don't, the gases in your lungs and digestive tract will expand causing your lungs to rupture and death to occur.

K

Beyond 90 seconds lung damage will become severe with major haemorrhaging and serious brain-damage occurring.

J

If re-compression occurs within 60 – 90 seconds, survival is possible. However, if heart action has stopped then death is inevitable. The damage to your lungs will be mild to moderate if re-compression occurs within 90 seconds.

E

Approaching 14 seconds, with the reduction in pressure, the boiling point of water drops and the water in your mouth will evaporate. If you are still conscious you'll feel a tingling sensation. Gas and water vapour continue to flow out of your mouth and nose reducing their temperature to near freezing.

I

After a minute the pressure in the veins exceeds the pressure in the arteries and blood circulation is effectively stopped.

F

By 15 seconds the oxygen-deprived blood will reach the brain. You will now be unconscious so your survival will depend on others helping you.

H

The nitrogen in the blood forms bubbles due to the low pressure.

D

Around ten seconds you will start to experience 'the bends'.

G

Water in your soft tissues evaporates causing you to swell to twice your size. If you survive this you should return to normal. Well-fitted elastic garments can help counter some of this swelling. There will probably be bruising but the skin is strong enough not to break so you won't burst.

C

Within five to eleven seconds you will start to lose consciousness so make it count. However, being more active uses up oxygen more quickly.

Launch sites

Rockets travel very fast, but they don't necessarily go upwards very far. After all, most satellites are only at altitudes of a few hundred kilometres, but they are travelling at over 20,000 kilometres per hour around the Earth. They also can't change direction very easily, so it's important that they are launched in the correct direction.

The Earth's equator is spinning at over 1,600 kilometres per hour in an easterly direction. Many launch sites are close to the equator, using this 'free' speed boost to reduce the amount of fuel required to reach equatorial orbits. To avoid the risk of damage to people or property, most launches take place over the sea.

United States

Cape Canaveral, home to the Kennedy Space Center, is arguably the most famous launch site in the world. While this makes it ideal for reaching the International Space Stations's orbit, it can't be used for polar orbits. Those are achieved from Vandenberg Air Force Base, in California. The US also operates a number of other launch sites around the world.

Carrier aircraft, US

Orbital Science's Pegasus rocket is suspended below a carrier aircraft, and can be used to launch small payloads in almost any direction.

Kodiak Island, US

Wallops Flight Facility, US

Vandenberg Air Force Base, US

Cape Canaveral, US

Carrier aircraft, US

Hammaguir, Algeria

Centre Spatial Guyanais, French Guiana

> 1,000 total launches

500

100

20

< 10

Number of launches per trajectory

100

50

20

< 5

> 100

Centre Spatial Guyanais

CSG has been Europe's launch site since 1970, and can launch into both equatorial and polar orbits. It replaced many European countries' own launch sites.

Russia

Russia doesn't have any coastal launch sites, but its rockets launch over uninhabited areas to the north. The first person in space, Yuri Gagarin, launched from Baikonur, Kazakhstan, which now launches all crewed missions to the International Space Station.

Plesetsk Cosmodrome, Russia

Dombarovsky Cosmodrome, Russia

Kapustin Yar Cosmodrome, Russia

Svobodny Cosmodrome, Russia

Tiuyuan Satellite Launch Center, China

Sohae Satellite Launching Station, North Korea

Naro Space Center, South Korea

Baikonur Cosmodrome, Kazakhstan

Semnan Space Center, Iran

Palmachim Airbase, Israel

Jiuquan Satellite Launch Center, China

Uchinoura Space Center, Japan

Tanegashima Space Center, Japan

Xichang Satellite Launch Center, China

Sriharikota Range, India

Reagan Test Site, Marshall Islands

EQUATOR

Broglio Space Centre / Kenya

China, Japan and India

Japan's space programme began in the 1960s, with China and India starting in the 1970s. With no coastal sites, China launches rockets over (relatively) uninhabited areas.

Woomera Range, Australia

Woomera

From 1969 – 1971, the UK used Woomera Range in Australia to launch Black Arrow rocket. There were only two successful launches before the project was cancelled, making the UK the only nation to have started and then cancelled a spaceflight programme.

Orbits

Satellites orbit the Earth at a range of distances, from just a few hundred kilometres above the surface to tens of thousands of kilometres away.

H ▬▮▬

(H) Highly-elliptical Orbit / 500 satellites

Some communications and astronomical satellites are in very elongated orbits, and can reach huge distances from Earth.

Altitude up to 100,000 km
Period / 2 – 20 hours

< Sun / 149,600,000 km away

▬▮▬ G

(G) Geostationary Orbit / 1,000 satellites

If they're at the right height above the equator, satellites orbit the Earth at the same rate as it rotates, and stay over the same spot on the surface.

Altitude / 36,000 km
Period / 23 hours 56 minutes

Some astronomy satellites go much further from Earth. Those observing the Sun often travel to the Lagrangian point one (L1), a gravitational sweetspot 1.5 million kilometres towards the Sun. Spacecraft observing the distant cosmos often travel to L2, which is a similar distance in the opposite direction. A few leave Earth orbit and trail either ahead or behind as they orbit the Sun, gradually drifting further away over time.

< Sun / 149,600,000 km away

Sun-Earth L1 / 3 satellites ▬▮▬

Altitude / 1,500,000 km
Period / 365.25 days

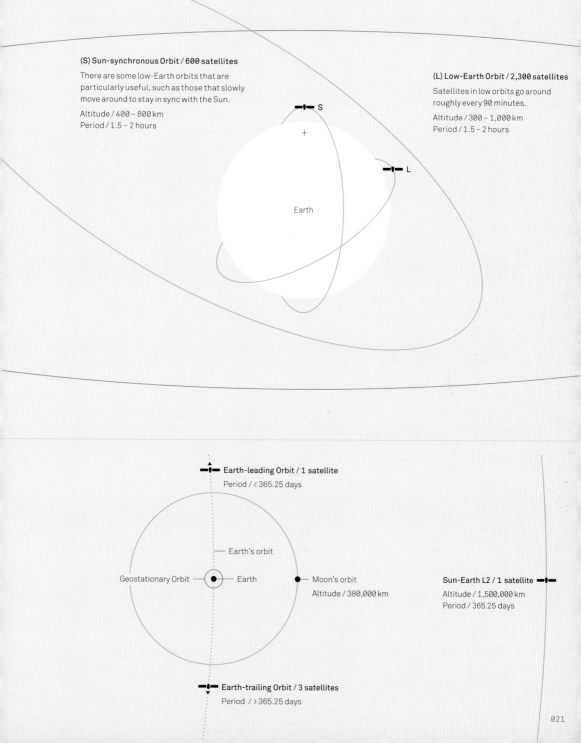

(S) Sun-synchronous Orbit / 600 satellites

There are some low-Earth orbits that are particularly useful, such as those that slowly move around to stay in sync with the Sun.

Altitude / 400 – 800 km
Period / 1.5 – 2 hours

S

(L) Low-Earth Orbit / 2,300 satellites

Satellites in low orbits go around roughly every 90 minutes.

Altitude / 300 – 1,000 km
Period / 1.5 – 2 hours

L

Earth

Earth-leading Orbit / 1 satellite
Period / < 365.25 days

Earth's orbit

Geostationary Orbit — Earth

Moon's orbit
Altitude / 380,000 km

Sun-Earth L2 / 1 satellite
Altitude / 1,500,000 km
Period / 365.25 days

Earth-trailing Orbit / 3 satellites
Period / > 365.25 days

Space junk

Each day tens of tonnes of rock fall from space.
Since 1957 this natural shower of meteors has been
joined by bits of rocket, dead satellites and even
entire space stations. Not all of these fall to Earth
though and some are left in orbit. Far from harmless,
our space age littering has left a dangerous legacy
as we surround our planet with debris travelling
at speeds of up to 28,000 kilometres per hour.

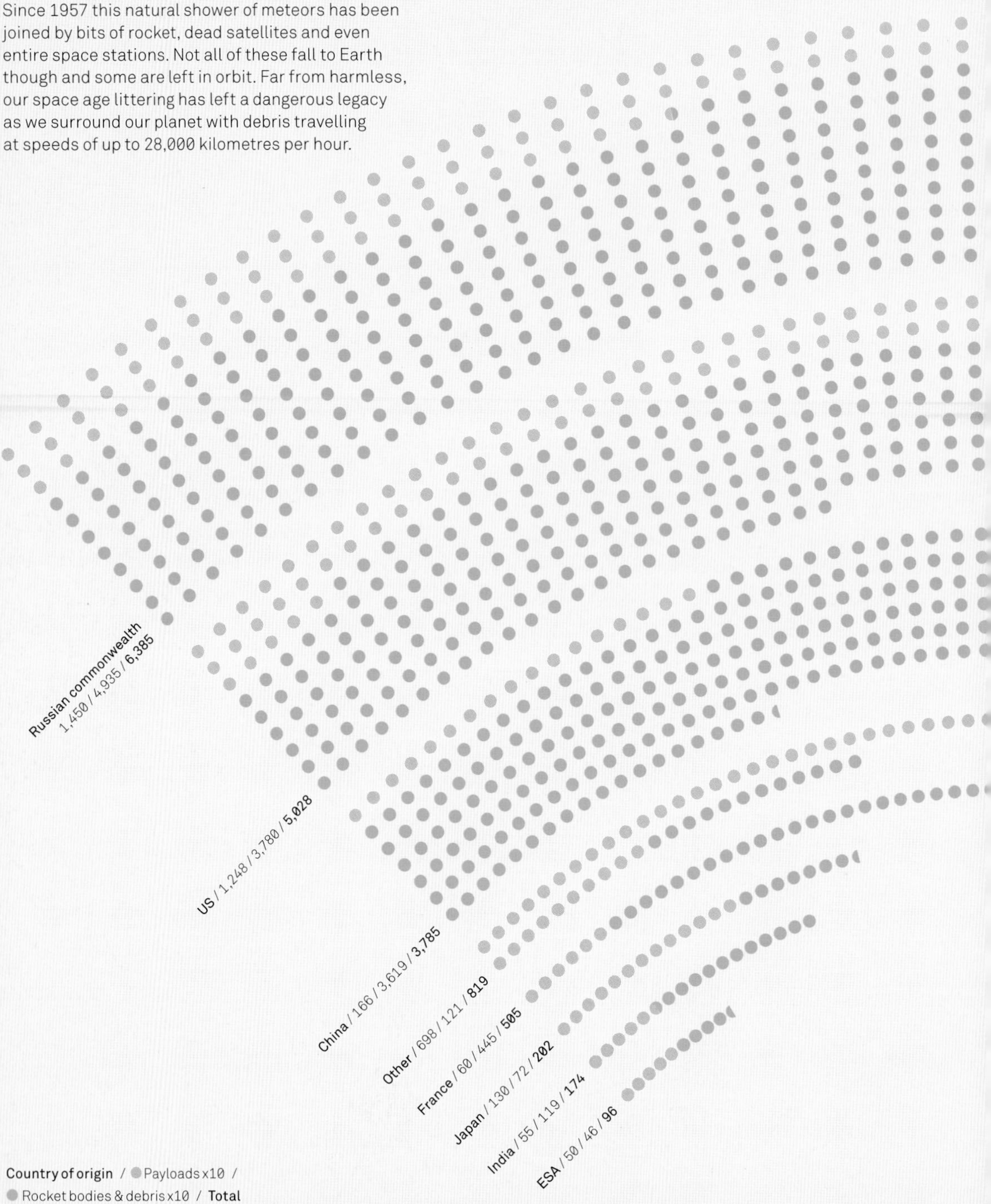

Russian commonwealth
1,450 / 4,935 / 6,385

US / 1,248 / 3,780 / 5,028

China / 166 / 3,619 / 3,785

Other / 698 / 121 / 819

France / 60 / 445 / 505

Japan / 130 / 72 / 202

India / 55 / 119 / 174

ESA / 50 / 46 / 96

Country of origin / ● Payloads x10 /
● Rocket bodies & debris x10 / **Total**

The most immediate threat from space junk is to the lives
of astronauts aboard the International Space Station and
the Tiangong Space Station. These 'hypervelocity impacts'
are a very real problem. In June 2014 a 10-centimetre puncture
hole was found on the ISS, extremely close to coolant tubes
in a photovoltaic radiator.

To tackle the issue, the world's major space agencies set up the
Inter-Agency Space Debris Coordination in 1993. As part of this,
NASA and the US Department of Defense track objects larger
than ten centimetres in Low-Earth Orbit and objects larger than
one metre in Geosynchronous Orbit.

Space stations

Getting to space is one thing. Staying there is much harder. You need supplies of oxygen to breathe, food to eat and you need to be able to deal with the waste products.

The first space station was Russia's Salyut 1 launched in 1971 with three compartments. The US put their first space station – Skylab – in orbit in 1973. It was visited by three crews and eventually came down over Australia in 1979.

During the 1970s, the Russians sent a series of replacement Salyuts into orbit (Salyuts 3 – 7). This set the stage for the much more ambitious Mir space station launched in 1986. Mir was continuously occupied for ten years, providing insight into the effects on the body of long-term human spaceflight.

In 1998, 16 nations started to construct the largest station yet: the International Space Station. It used a modular system that allowed different parts to be made by different countries and connected together. It now consists of 14 pressurised modules covering a range of purposes. It has been continuously crewed since 2000. The Chinese launched the Tiangong-1 space station in 2011 and plan to launch a larger station in the 2020s.

↟ ↟ ↟
Tiangong-1, Chinese
Operational lifetime / 2011 – 2020 estimate
Length / 10.4m

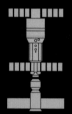

↟ ↟ ↟
Salyut (1, 3 – 7) / Soyuz capsule, Russia
Operational lifetime /1971 – 1991
Length / 15.8m

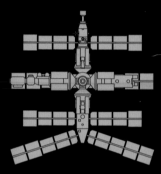

↟ ↟ ↟ ↟ ↟ ↟
Mir space station, Russia
Operational lifetime / 1986– 2001
Length / 31m

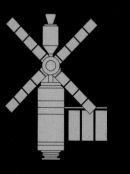

↟ ↟ ↟
Skylab, NASA
Operational lifetime /1973 – 1979
Length / 26.3m

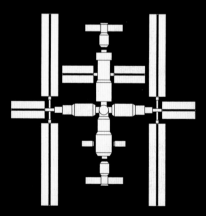

↟ ↟ ↟
Chinese space station (planned)
Operational lifetime / 2023–
Length / 20m

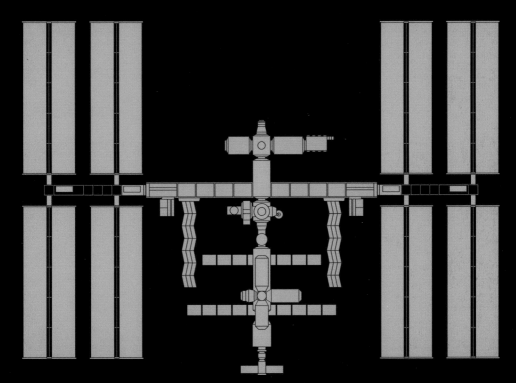

International Space Station (ISS)
Operational lifetime / 1998– 2024 estimate
Length / 109m

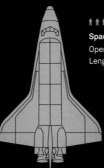

Space Shuttle (retired)
Operational lifetime /1981 – 2011
Length / 37 m

Journey to the Moon

On 16 July 1969, a Saturn V rocket launched from Kennedy Space Center bound for the Moon. The crew of three would traverse the 380,000 kilometres between the Earth and Moon over the next three days.

Once in lunar orbit, Neil Armstrong and Edwin 'Buzz' Aldrin left Michael Collins and took the Eagle Lunar Module down to the surface. The final moments of descent required their piloting expertise as they had to avoid a boulder field. They landed with only 25 seconds of fuel remaining.

After stepping out onto the Moon at 02:56 UTC on 21 July, Aldrin and Armstrong performed various tests, collected soil samples, took many photographs and spoke to President Nixon. They returned to the Lunar Module about two-and-a-half-hours later. Following several hours of rest, they prepared for lift-off at 17:54 UTC. They rejoined Michael Collins in orbit and then started the journey back to Earth.

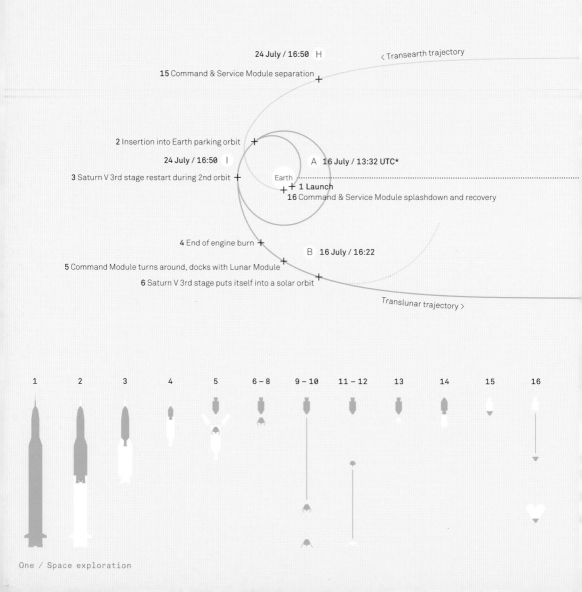

24 July / 16:50 H   < Transearth trajectory

15 Command & Service Module separation

2 Insertion into Earth parking orbit

24 July / 16:50 I

A 16 July / 13:32 UTC*

3 Saturn V 3rd stage restart during 2nd orbit

Earth

1 Launch

16 Command & Service Module splashdown and recovery

4 End of engine burn

B 16 July / 16:22

5 Command Module turns around, docks with Lunar Module

6 Saturn V 3rd stage puts itself into a solar orbit

Translunar trajectory >

1   2   3   4   5   6 – 8   9 – 10   11 – 12   13   14   15   16

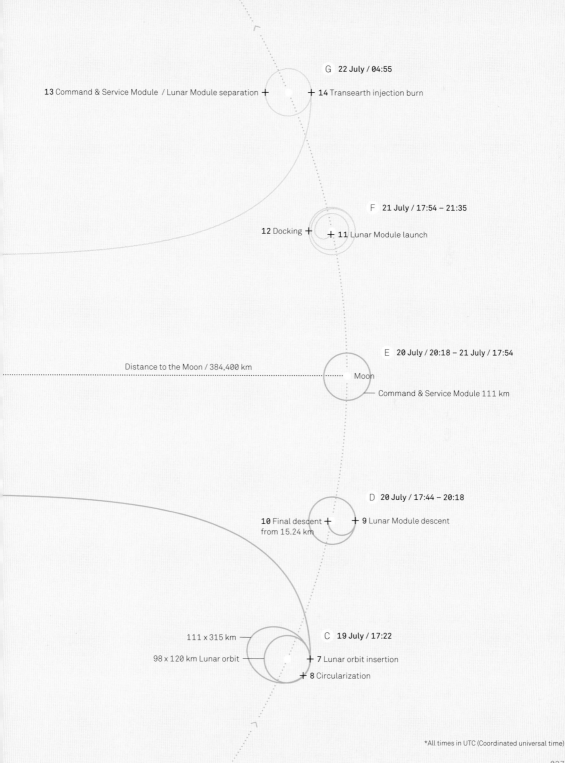

G **22 July / 04:55**

13 Command & Service Module / Lunar Module separation ✛   ✛ **14** Transearth injection burn

F **21 July / 17:54 – 21:35**

12 Docking ✛   ✛ **11** Lunar Module launch

E **20 July / 20:18 – 21 July / 17:54**

Distance to the Moon / 384,400 km

Moon

Command & Service Module 111 km

D **20 July / 17:44 – 20:18**

10 Final descent ✛   ✛ **9** Lunar Module descent
from 15.24 km

111 x 315 km

C **19 July / 17:22**

98 x 120 km Lunar orbit   ✛ **7** Lunar orbit insertion

✛ **8** Circularization

*All times in UTC (Coordinated universal time)

On the Moon

On 20 July 1969 at 20:18 UTC, the *Apollo 11* Lunar Module (LM) touched down on the surface of the Moon. They had only 45 seconds of fuel left in the tank.

Astronauts Neil Armstrong and Buzz Aldrin spent two hours preparing the LM so it would be ready to take off when necessary. They then had a meal break and prepared for their walk outside. About six and a half hours after landing, Neil Armstrong stepped out of the hatch to the top of the ladder. He swung out a black and white TV camera to film the event and descended the ladder. On the surface he said his famous lines and was then joined by Buzz Aldrin.

They placed a TV camera some distance from the LM, planted a US flag and talked to the US President. After some experimentation with the best method of moving around, they explored the local area, took photographs, set up a few experiments, and collected rock samples to bring back. In total they brought back around 20 kilogram of lunar material.

After a relatively brief 21.6 hours on the lunar surface, the first visitors from planet Earth launched themselves back into orbit. Five more Apollo missions followed.

15 ● ● 17
12 ●● 14 ● 11
16
Apollo Lunar landing sites

Apollo 11 landing site
0.67409 degrees north / 23.47298 degrees east

105 m

69 m

Lunar craters

TV camera ◀

✳ P3

Solar wind composition experiment
Lunar surface closeup camera
Flag
Footprint ✎
Lunar module
P2 ✳

✳ Panorama 1

Little west crater

Double crater

✳ P4

P5 ✳

— Area covered by astronauts

Laser ranging experiment ■

● Discarded cover

Ⱶ Passive seismic experiment

Football pitch to scale / Wembley Stadium

Equipment left on the Moon by Apollo 11

LM descent stage
Gold olive branch
Apollo 1 patch
Cosmonaut medals
Moon memorial disk
TV camera
TV subsystem
TV wide-angle lens
TV day lens
TV cable assembly (30.5 m)

Polarizing filter
S-band antenna
S-band antenna cable
Flag kit
Experimental central station
Passive seismic experiment
Laser range reflector
Portable Life Support System
Oxygen filter
Remote control unit

Urine collection assemblies
Defecation collection device
Overshoes
Bags
Gas connector covers
Waist tether
Life line
Conveyor assembly
Food assembly (4 man days)
SRC/OPS adapter

Canister, ECS LIOR
Containers
Pallet assembly x2
Primary structure assembly
Hammer
Large sample scoop
Extension handle
Tongs
Gnomon (excluding mount)
Ascent stage

Apollo 11 Lunar Module

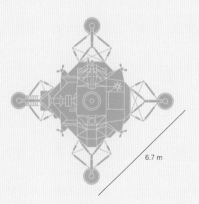

6.7 m

2.8 m

6 m

9.4 m

3.2 m

Missions to the Moon

For millennia humans have looked at the Moon. Some have dreamed of visiting our celestial neighbour. In the 20th Century those dreams were finally turned into reality as both the USSR and US built rockets that were capable of reaching across the void.

The early missions were fraught with difficultly and many exploded at launch or shortly afterwards. Others simply failed to move to the correct orbit. The first confirmed success was the USSR's *Luna 2* in 1959, followed by *Luna 3* a few weeks later. There followed a string of failures until the first successful American mission *Ranger 7* in 1964. Over the next five years, reliability improved and the first humans orbited the Moon during Christmas 1968 (just three months after *Zond 5* took a tortoise there and back). Overall, the US took 12 people to the surface of the Moon and back.

After the Apollo missions, interest in the Moon waned for many years. The Millennium saw a renewed space race including missions from Europe, India and China that all explored the Moon.

— Successful missions
   Failed missions

A / **Able I** 1958
B / **Luna 2** 1959
C / **Ranger 7** 1964
D / **Luna 9** 1966
E / **Apollo 8** 1968
F / **Apollo 11** 1969
G / **Apollo 13** 1970 (partial failure)
H / **Luna 17** 1970
I / **Apollo 17** 1972
J / **Luna 21** 1973
K / **Luna 24** 1976
L / **Hiten** 1990
M / **SMART-1** 2003
N / **Chandrayaan-1** 2008
O / **Chang'e 3** 2013

Earth

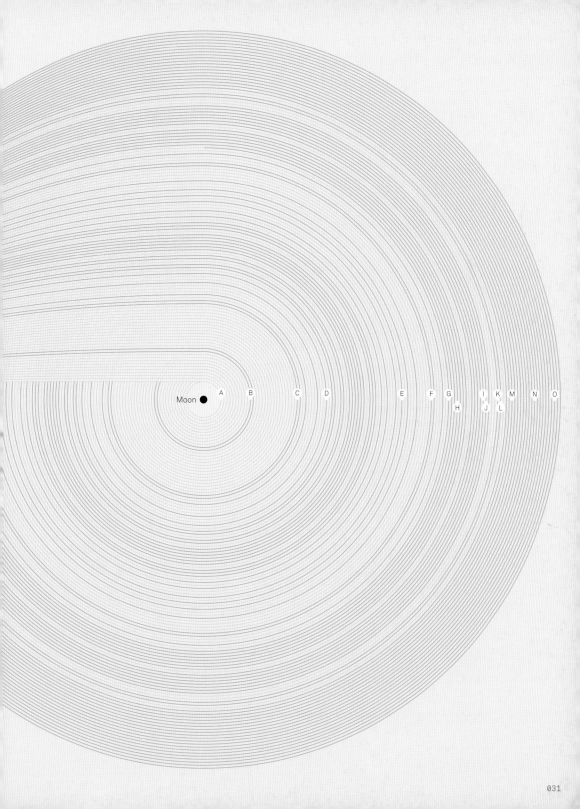

Moon ● A B C D E F G I K M N O
H J L

Interplanetary missions

Space exploration is difficult. Since the 1960s there have been over a hundred missions to about 30 different bodies in the Solar System – and that's not including the Moon.

Not all have succeeded, with many early missions to Mars and Venus failing before reaching their destination. Some spacecraft have visited more than one object en route. Others, such as *Pioneers 10 & 11* and *Voyagers 1 & 2*, have kept on travelling out of the Solar System.

By the time you read this, many of the ongoing mission may have reached their destinations: Juno to Jupiter, Akatsuki to Venus, and *Hayabusa 2* to *1999 JU3*.

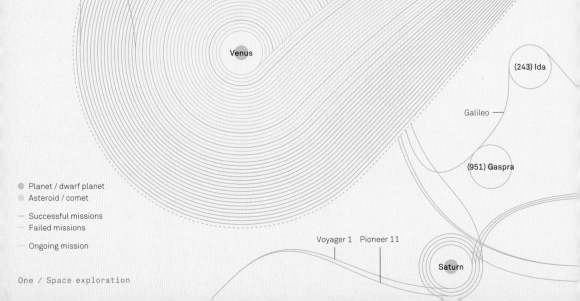

Mars

Mercury

Venus

(243) Ida

Galileo —

(951) Gaspra

● Planet / dwarf planet
● Asteroid / comet

— Successful missions
— Failed missions

⋯ Ongoing mission

Voyager 1   Pioneer 11

Saturn

Ceres

67P/Churyumov-Gerasimenko

Dawn

Rosetta

(21) Lutetia

Vesta

Deep Impact

103P/Hartley

(2867) Šteins

9P/Tempel

Stardust

(162173) 1999 JU3

81P/Wild

Hayabusa 2

Stardust NExT

(25143) Itokawa

(5535) Annefrank

Hayabusa

(433) Eros

CONTOUR

Earth

2P/Encke

73P/Schwassmann-Wachmann

21P/Giacobini-Zinner

1P/Halley

Chang'e 2

ICE

Giotto

(4179) Toutatis

Deep Space 1

26P/Grigg-Skjellerup

(9969) Braille

Pluto

(132524) APL

New Horizons

Pioneer 10

19P/Borrelly

Jupiter

Voyager 2

Uranus

Vega 1 & 2

Neptune

Distant voyagers

MESSENGER / 90.4 Astronomical Units

NASA's MESSENGER mission launched in 2004 on a trip to the innermost planet, Mercury. Getting so close to the Sun and into an orbit around Mercury requires a large loss of speed. A long trip with visits to Earth, Venus and Mercury was needed to use their gravity to slow the probe down.

Voyagers, Pioneers and New Horizons

NASA's *Pioneers 10 & 11* launched in 1972 and 1973 were the first missions to visit Jupiter. In 1977 NASA launched *Voyagers 1 & 2* to visit Jupiter and Saturn as they were favourably aligned. *Voyager 2* carried on to be the only probe to visit Uranus and Neptune. In 2006 New Horizons became the first probe designed to visit Pluto and reached it in July 2015.

MESSENGER / plan

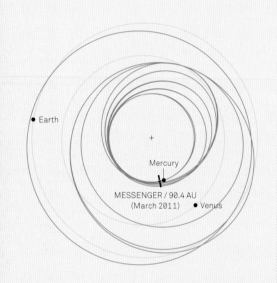

Voyagers, Pioneers and New Horizons / plan

- ● New Horizons / 35.0 AU (July 2015)
- ● Voyager 1 / 142.3 AU
- ● Voyager 2 / 131.6 AU
- ● Pioneer 10 / 122.6 AU
- ● Pioneer 11 / 118.6 AU

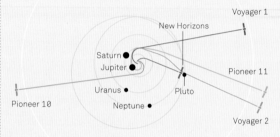

MESSENGER / elevation

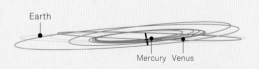

Voyagers, Pioneers and New Horizons / elevation

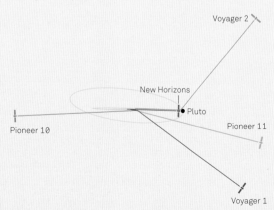

1 Astronomical Unit (AU) = 149.6 million km (distance from the Earth to the Sun)

Rosetta / 42.8 Astronomical Units

ESA's Rosetta mission launched in 2004 and spent
10 years chasing down comet *67P/Churyumov–
Gerasimenko*. On the way it visited Mars, Earth and
two asteroids. In late 2014 Rosetta became the first
spacecraft to orbit a comet nucleus and also the first
to land a probe on a comet surface successfully.

Ulysses / 79.2 Astronomical Units

This NASA/ESA probe launched in 1990 on a mission
to observe the Sun from a range of angles. To reach an
orbit that took it up and over the poles of the Sun required
using the gravity of the planet Jupiter to hurl it out of the
plane of the Solar System.

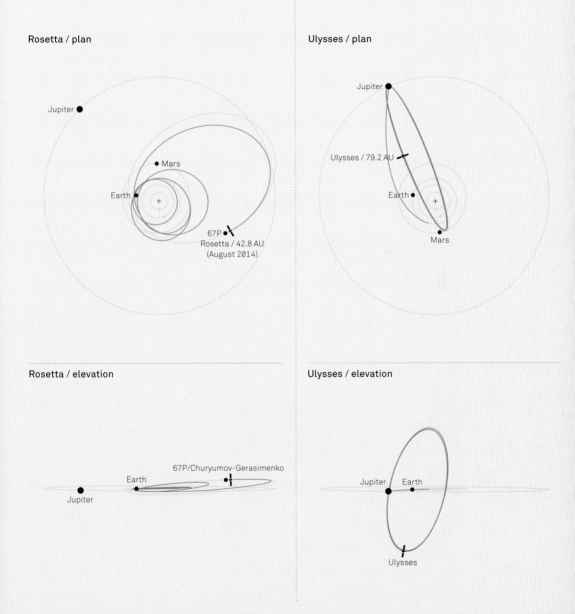

Rosetta / plan

Jupiter

Mars

Earth

67P
Rosetta / 42.8 AU
(August 2014)

Ulysses / plan

Jupiter

Ulysses / 79.2 AU

Earth

Mars

Rosetta / elevation

67P/Churyumov-Gerasimenko

Earth

Jupiter

Ulysses / elevation

Jupiter   Earth

Ulysses

Planetary rovers

By 1970 the race to the Moon was over. Now it was time to explore the surface. The Soviet Union leapt into the lead with robotic roving, sending two Lunokhod rovers to the Moon. *Lunokhod 2's* distance record was held for decades.

NASA equipped the later Apollo astronauts with lunar rovers, lightweight buggies that allowed them to travel much futher and faster than the earlier pioneers.

NASA's first robotic rover – the teatray-sized rover Sojourner – went to the red planet. Despite only travelling around 100 metres, and straying no more than 12 metres from the main lander, it led the way for the next generation.

Spirit and Opportunity landed on Mars in 2004, with an intended mission duration of 90 days. Both far exceeded the expectations of their human masters. Spirit got stuck in soft sand in 2009 and succumbed to the cold Martian winter in 2010.

Opportunity was still going a decade after landing, and broke the distance record for extra-terrestrial roving in 2014. These rovers confirmed that billions of years ago Mars was a much warmer, wetter world.

The most advanced rover to date is Curiosity, a car-sized rover which landed in 2012. With ten scientific instruments

~~ Moon   ~~ Mars

Lunokhod 1 CCCP / Moon 1970 – 1971   10.54 km

Apollo 15 NASA / Moon 1971

Apollo 16 NASA / Moon 1972

Apollo 17 NASA / Moon 1972

Lunokhod 2 CCCP / Moon 1973

Sojourner NASA / Mars 1997 **100 m**

Spirit NASA / Mars 2004 – 2010   7.73 km

Opportunity NASA / Mars 2004 –

Curiosity NASA / Mars 2012 –   10.30 km*

Yutu CNSR / Moon 2012 **40 m**

on board, including a laser, it has shown that Mars was once capable of supporting life. The question of whether there ever was life there, however, will be left to future explorers, be they robotic or human, to answer.

In 2013, China successfully landed Chang'e 3 on the Moon and with it the Yutu rover. Yutu circled the lander and traversed a further 40 metres before a mechanical failure prevented it from roving, though it continued to send panoramic photos for months afterwards.

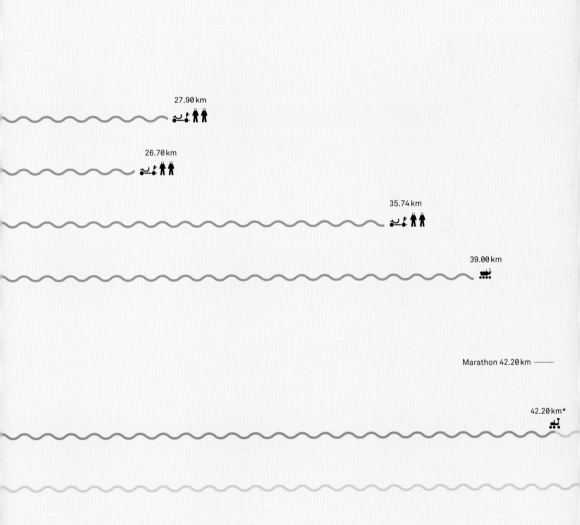

27.90 km

26.70 km

35.74 km

39.00 km

Marathon 42.20 km ———

42.20 km*

* Distance travelled as of March 2015

Two / Solar System

How many planets?

Planet comes from the ancient Greek meaning 'wanderer' and included any celestial object that appeared to move compared to the stars. For millenia the planets were known to be Mercury, Venus, Mars, Jupiter, Saturn, Sun, and Moon. That all changed in the 17th Century when the discovery of objects near Jupiter and Saturn swelled the ranks to 16.

As our understanding of the Solar System improved and, with observations showing comets to be celestial, the definition of what was a planet started to change for the first time. A planet became an object that orbited the Sun with a nearly circular path. The Sun was no longer a planet. The Earth was. The Moon, along with the satellites of Jupiter and Saturn, moved into a new class of moons.

Under the new definition, the discovery of Uranus became the first new planet since antiquity. In the early 19th Century numbers swelled once more when Neptune was discovered, along with Ceres, Pallas, Juno, Vesta, Hencke, taking the total to 13. In 1847, these small planets orbiting between Mars and Jupiter were reclassified as asteroids dropping the number of major planets back down to eight for the third time.

In 1930, the discovery of Pluto took us back to nine. In the 21st Century, Haumea, Eris, and Makemake were found and led to a redefinition of what makes a planet and created a new class of dwarf planet. Our Solar System now contains eight planets, five dwarf planets, 182 moons, and over 650,000 asteroids. For now...

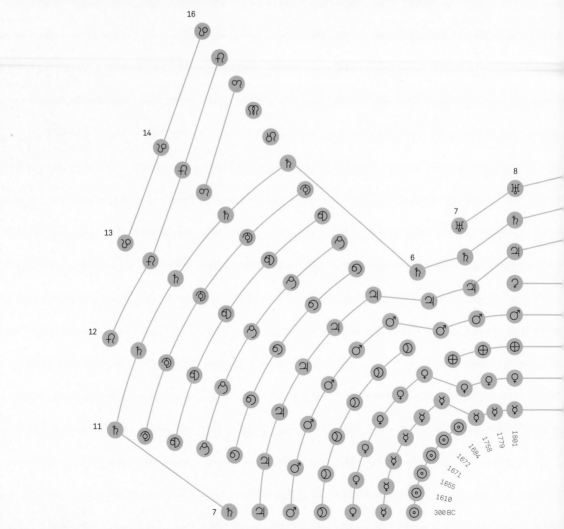

Planet classification change over time

- ⬤ Planet
- ⬤ Dwarf planet

♇ Astraea	♠ Haumea	☽ Moon	♁ Titan
◉ Callisto	⚵ Iapetus	♆ Neptune	⛢ Uranus
⚳ Ceres	⚲ Io	⚴ Pallas	♀ Venus
♏ Dione	⚹ Juno	♇ Pluto	⚶ Vesta
⊕ Earth	♃ Jupiter	♍ Rhea	
⚸ Eris	♏ Makemake	♄ Saturn	
♋ Europa	♂ Mars	☉ Sun	
♋ Ganymede	☿ Mercury	♌ Tethys	

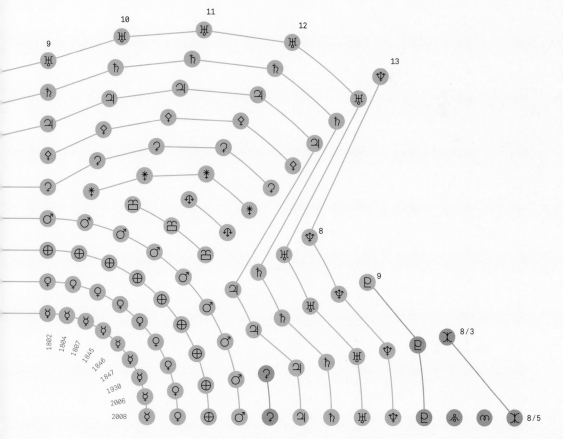

Scale model Solar System

Imagining the distances to the planets can be tricky,
not least because they're so vast relative to their
size. If the Sun were moved to Paris and shrunk
to about the height of the Eiffel Tower, Mercury
would skirt the outer suburbs of Paris. Meanwhile,
the Earth would be nearly 40 kilometres away, and
the size of an African elephant. While Jupiter would
stay largely within France, Saturn's orbit would run
through Brussels and London, and Uranus would
pass through Munich and Liverpool. The outermost-
planet, Neptune, could be found near Copenhagen,
while the Kuiper Belt objects would be holidaying
in Morocco or the Azores.

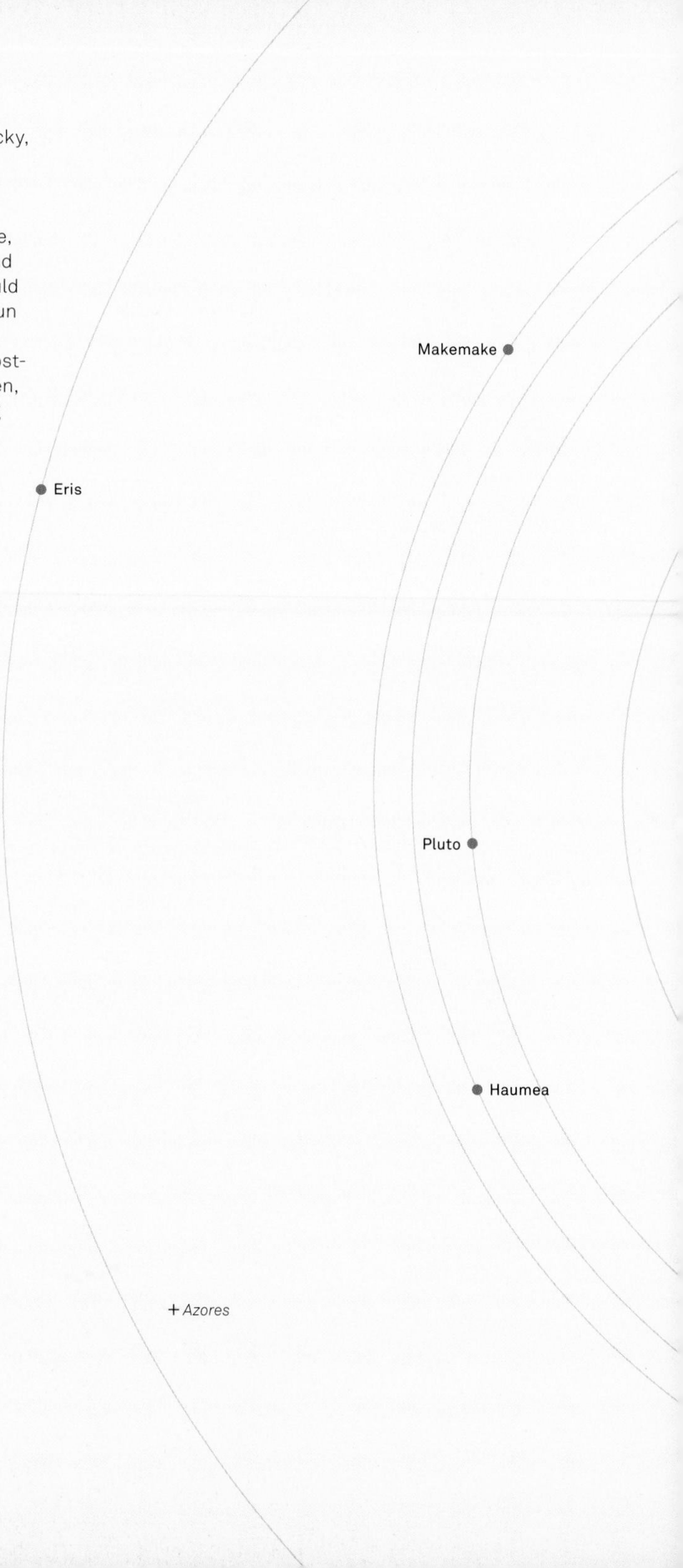

Makemake ●

● Eris

Pluto ●

● Haumea

+ *Azores*

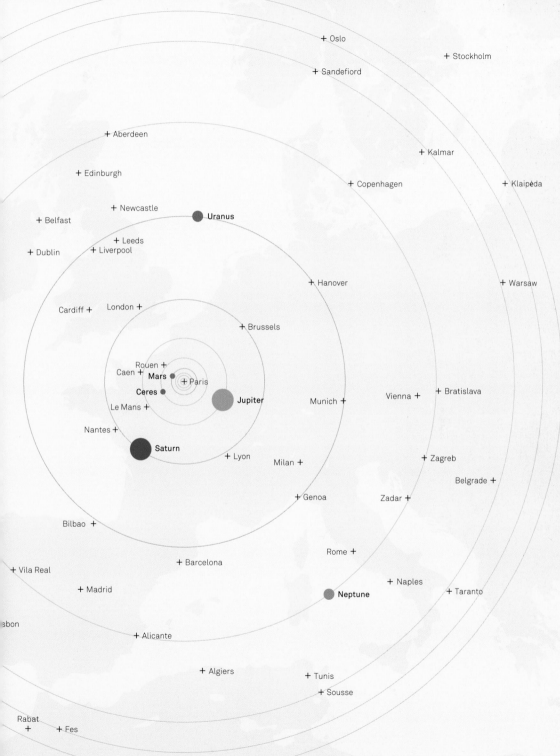

+ Oslo
+ Stockholm
+ Sandefjord
+ Aberdeen
+ Kalmar
+ Edinburgh
+ Copenhagen
+ Klaipėda
+ Newcastle
● **Uranus**
+ Belfast
+ Leeds
+ Liverpool
+ Dublin
+ Hanover
+ Warsaw
Cardiff +
London +
+ Brussels
Rouen +
Caen +
Mars ●
Ceres ●
+ Paris
● **Jupiter**
Munich +
Vienna +
+ Bratislava
Le Mans +
Nantes +
● **Saturn**
+ Lyon
Milan +
+ Zagreb
Belgrade +
Genoa +
Zadar +
Bilbao +
Rome +
+ Vila Real
+ Barcelona
+ Naples
● **Neptune**
+ Taranto
+ Madrid
sbon
+ Alicante
+ Algiers
+ Tunis
+ Sousse
Rabat
+
+ Fes

043

The family of planets

The constituents of our Solar System vary hugely in size from small asteroids only kilometres across to the gigantic Jupiter, nearly 140,000 kilometres in diameter.

The eight planets fall into three main categories. The four nearest the Sun are comprised mostly of rock. The two largest planets, Jupiter and Saturn, are the gas giants, made almost entirely of hydrogen and helium. The two most distant planets, the ice giants Uranus and Neptune, have larger solid cores and methane clouds in their atmospheres. Most of the dwarf planets lie in the Kuiper Belt, and are the largest examples of a collection of icy bodies that orbit out beyond Neptune. At such great distances these are hard to observe, and there are almost certainly many objects in the Kuiper Belt which are still to be discovered.

Many of the Solar System's smaller objects orbit in the asteroid belt, between Mars and Jupiter. The largest of these by far is Ceres, though there are dozens more that measure hundreds of kilometres in diameter. The four biggest objects in the asteroid belt make uphalf its total mass.

Venus
12,104 km
0.615 years
5,832* hours

Mars
6,792 km
1.881 years
24.6 hours

Mercury
4,879 km
0.2 years
1,408 hours

Earth
12,756 km
1 year
24.0 hours

Ceres (D)
952.4 km
4.6 years
9.8 hours

Day length / 9.9 hours

Jupiter
139,822 km

Orbital period / 11.9 Earth years

Planet name (D = dwarf planet)
Diameter
Orbital period, Earth years
Day length

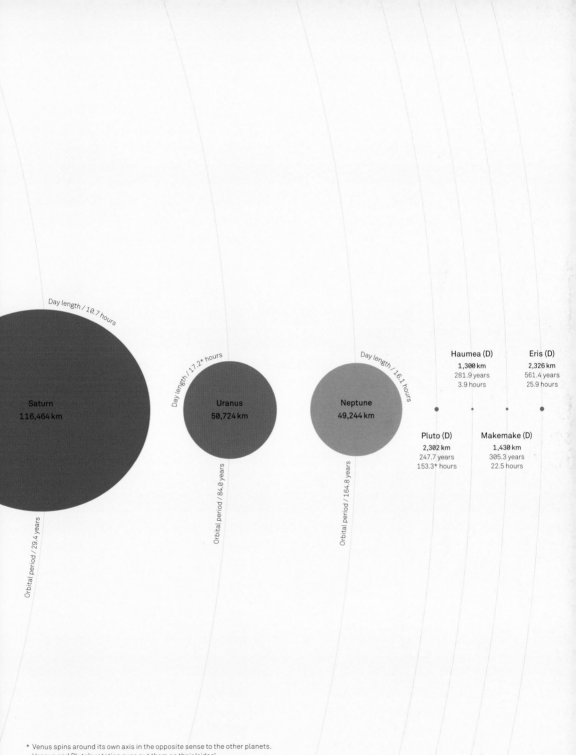

Day length / 10.7 hours

Saturn
116,464 km

Orbital period / 29.4 years

Day length / 17.2* hours

Uranus
50,724 km

Orbital period / 84.0 years

Day length / 16.1 hours

Neptune
49,244 km

Orbital period / 164.8 years

Haumea (D)
1,300 km
281.9 years
3.9 hours

Eris (D)
2,326 km
561.4 years
25.9 hours

Pluto (D)
2,302 km
247.7 years
153.3* hours

Makemake (D)
1,430 km
305.3 years
22.5 hours

* Venus spins around its own axis in the opposite sense to the other planets.
  Uranus and Pluto's rotation axes put them on their 'sides'.

Moons

Today we think of a moon as a celestial body that orbits a planet. The most obvious example is our nearest neighbour – the Moon – which orbits Earth. It wasn't until January 1610 that we first found moons around another planet when Galileo turned his telescope towards Jupiter. The first of Saturn's moons was discovered in 1655. Uranus' first moons were seen in 1851 – 70 years after the planet itself was discovered. We had to wait until as late as 1877 before Mars' two tiny moons were finally identified.

The 20th and 21st Centuries saw a huge increase in the number of known moons thanks, in part, to advances in telescope resolution but mostly from the fleet of spacecraft we've sent out to visit the planets.

We now know of moons around Earth, Mars, Jupiter, Saturn, Uranus, Neptune, Pluto, Eris and Haumea.

Naming

Historically, moons were often named by their discoverer but, since 1975, the International Astronomical Union has overseen the naming process. There are now several naming conventions depending on the parent body. The two moons of Mars are named after the sons of Ares (the Greek equivalent of Mars). Moons of Jupiter are named after the lovers or descendents of Jupiter (Zeus). Saturn's moons are named after giants or their descendents (currently from Greek, Norse, Gallic and Inuit mythologies). The moons of Uranus are named after characters in Shakespearean plays. Neptune's moons are named after Greek sea gods and Pluto's moons all have connections to Hades.

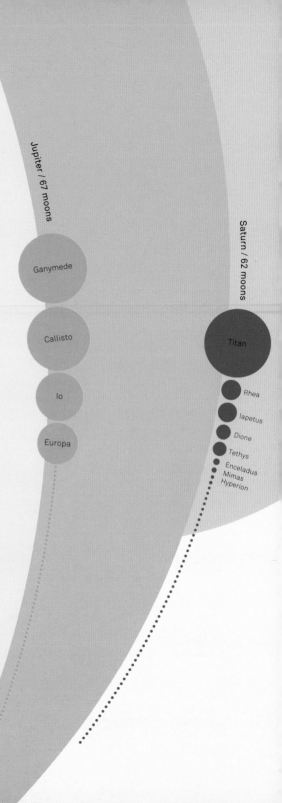

Jupiter / 67 moons

Ganymede

Callisto

Io

Europa

Saturn / 62 moons

Titan

Rhea

Iapetus

Dione

Tethys

Enceladus
Mimas
Hyperion

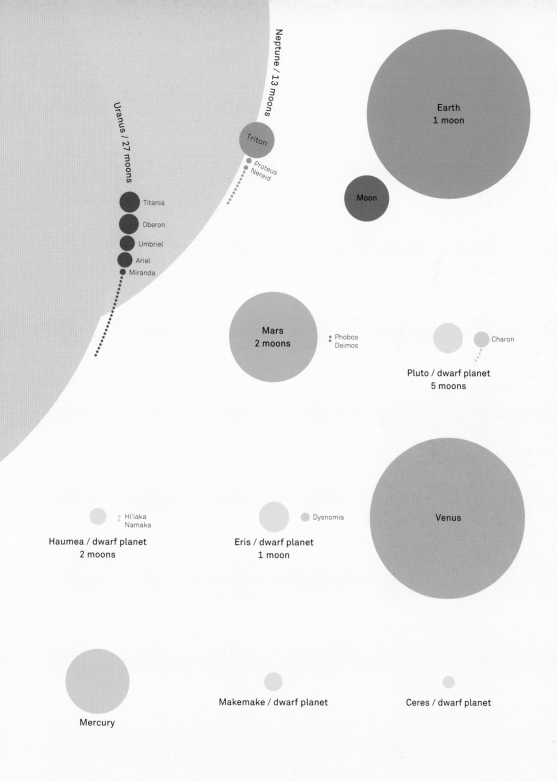

Eclipses

If you've ever had the opportunity to see a total solar eclipse you will know how special they are. Solar eclipses occur when the Moon passes between us and the disk of the Sun. Lunar eclipses happen when the Earth passes between the Moon and the Sun. But the Moon orbits the Earth every four weeks so why don't eclipses happen every fortnight?

The Moon's orbit is slightly tilted relative the Earth's so most of the time the Moon will appear to pass above or below the Sun and above or below the Earth's shadow.

Because its orbit is not circular, sometimes the Moon appears smaller and doesn't fully block the Sun. These solar eclipses are called 'annular eclipses' because a ring, or annulus, of sunlight remains visible.

The complicated movement of the Moon leads to many cycles in the frequency of eclipses. Amongst these are the 'semester' (177.2 days, or just under 6 months), the lunar year (354.4 days), and the 'saros' (6,585.32 days or just over 18 years).

● Total solar eclipse   ◖ Partial solar eclipse   ○ Annular solar eclipse   ◓ Hybrid solar eclipse
● Total lunar eclipse   ◖ Partial lunar eclipse   ○ Penumbral lunar eclipse

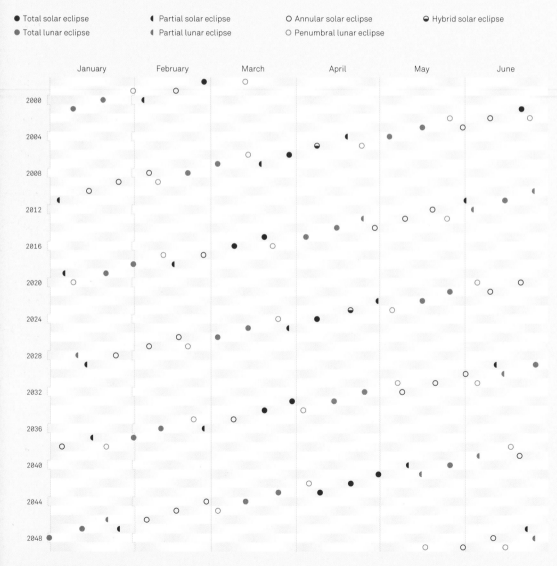

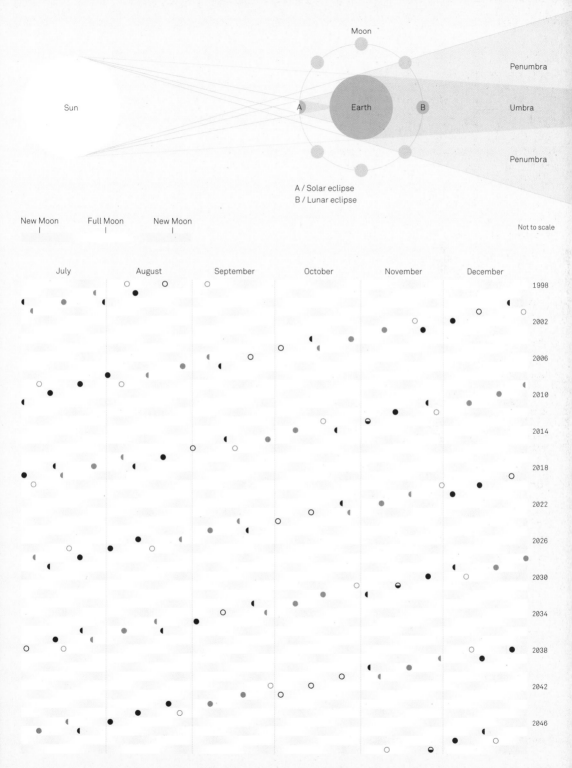

Moon

Sun

Earth

A

B

Penumbra

Umbra

Penumbra

A / Solar eclipse
B / Lunar eclipse

New Moon   Full Moon   New Moon

Not to scale

July   August   September   October   November   December

1998
2002
2006
2010
2014
2018
2022
2026
2030
2034
2038
2042
2046

Solar System top ten

Mountains

Heights of mountains in the Solar System from base to peak.
The largest is Vesta's South Pole Mountain discovered
in 2011 by the DAWN spacecraft.

10.2 km — Earth / Mauna Kea
829.8 m — Burj Khalifa / UAE
11.7 km — Mars / Arsia Mons
12.6 km — Mars / Elysium Mons
12.7 km — Io / Ionian Mons east ridge
13.4 km — Io / Euboea Montes

Lakes

The largest known lake – Kraken Mare – was found by the Cassini
spacecraft in 2007 and is composed of liquid hydrocarbons.

31,500 km² — Earth / Lake Baikal
32,000 km² — Io / Loki Patera (Lava)
32,893 km² — Earth / Lake Tanganyika
58,000 km² — Earth / Lake Michigan
59,600 km² — Earth / Lake Huron

Canyons, outflow channels and chasmata

The longest channel in the Solar System – Baltis Vallis – was
discovered by the *Venera 15 & 16* spacecraft. It probably once
held a river of lava.

740 km — Rhea / Galunlati Chasmata
1,219 km — Tethys / Ithaca Chasma
1,758 km — Mars / Ares Valles
3,160 km — Venus / Citlalpul Vallis
700 km — Venus / Ahsabkab Vallis
750 km — Earth / Grand Canyon of Greenland
1,580 km — Mars / Kasei Valles
1,720 km — Mars / Tiu Valles

Craters

Borealis Basin covers almost the entire northern hemisphere
of Mars. Its origin is uncertain but is thought to be the result
of an impact at some point in Mars' past.

505 km — Vesta / Rheasilvia
580 km — Iapetus / Turgis
715 km — Mercury / Rembrandt
1,145 km — Moon / Mare Imbrium
1,550 km — Mercury / Caloris Basin
2,300 km — Mars / Hellas Planitia

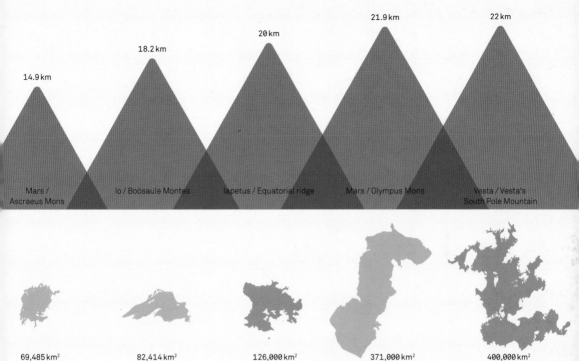

14.9 km
Mars /
Ascraeus Mons

18.2 km
Io / Boosaule Montes

20 km
Iapetus / Equatorial ridge

21.9 km
Mars / Olympus Mons

22 km
Vesta / Vesta's
South Pole Mountain

69,485 km²
Earth / Lake Victoria

82,414 km²
Earth / Lake Superior

126,000 km²
Titan / Ligeia Mare (Hydrocarbon)

371,000 km²
Earth / Caspian Sea

400,000 km²
Titan / Kraken Mare (Hydrocarbon)

6,800 km
Venus / Baltis Vallis

3,769 km
Mars / Valles Marineris

2,500 km
Moon / South
Pole–Aitken basin

3,000 km
Moon /
Procellarum Basin

3,300 km
Mars /
Utopia Planitia

8,500 km
Mars /
Borealis Basin

Structure of planets and moons

We know about the interior of the Earth thanks to seismology. Experiments left by lunar missions have used moon-quakes to understand the interior of the Moon. For the other bodies in the Solar System it's much harder to tell. Flybys of spacecraft, combined with physical models, can be used to estimate the interior structure.

The strong magnetic fields of Jupiter and Saturn mean they must have a conductive interior, probably made of 'metallic hydrogen'. Whether they have solid cores at their centres is much less clear. Uranus and Neptune are thought to have icy, rocky cores surrounded by a thick layer of ices, predominently water and methane.

Many moons in the outer Solar System are now thought to have sub-surface oceans of liquid water, due to internal heating caused by tides from their parent planets. This makes these mini-worlds just as exciting and mysterious as the planets they orbit.

- ⬤ Water
- ⬤ Ices
- ⬤ Rock / ice
- ⬤ Methane
- ⬤ Metallic hydrogen
- ⬤ Atmosphere
- ⬤ Molten rock
- ⬤ Rock
- ⬤ Molten iron
- ⬤ Solid iron
- ⬤ Solid iron (FeS-rich)

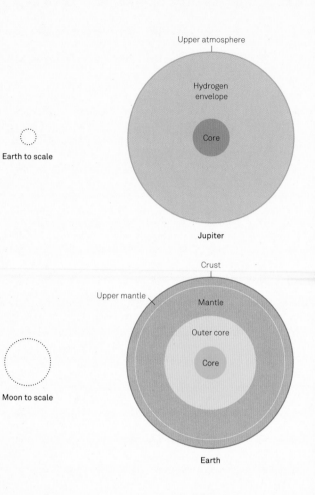

Earth to scale

Moon to scale

Jupiter

Earth

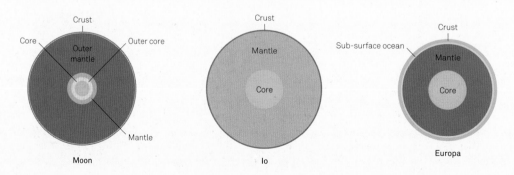

Moon

Io

Europa

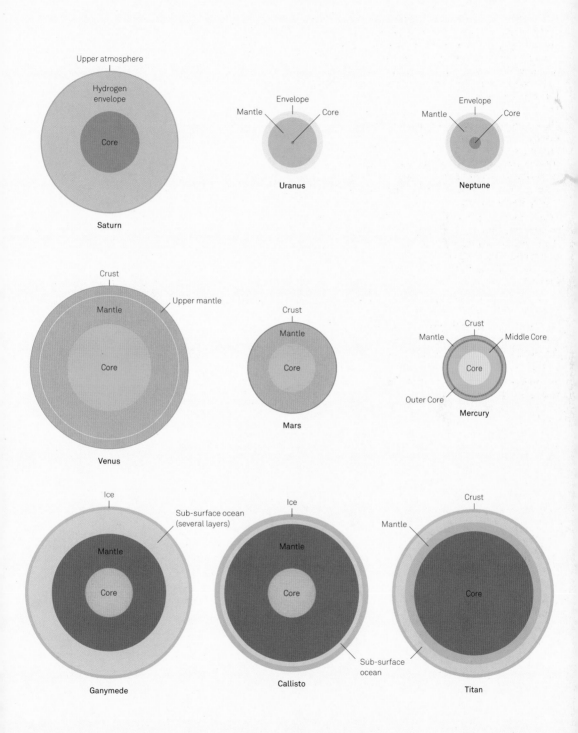

Upper atmosphere

Hydrogen envelope

Core

Saturn

Envelope

Mantle

Core

Uranus

Envelope

Mantle

Core

Neptune

Crust

Mantle

Upper mantle

Core

Venus

Crust

Mantle

Core

Mars

Crust

Mantle

Core

Middle Core

Outer Core

Mercury

Ice

Mantle

Sub-surface ocean (several layers)

Core

Ganymede

Ice

Mantle

Core

Sub-surface ocean

Callisto

Crust

Mantle

Core

Titan

Planetary atmospheres

The Earth's atmosphere is a thin protective layer which cushions us from the harshness of outer space. The coldest place in the atmosphere is the tropopause, where the thin air is insulated from the ground heat. This region is also protected from sunlight, which heats the upper layers of the atmosphere.

Venus' crushing pressures and scorching temperatures make its surface uninhabitable. At an altitude of around 50 kilometres, however, the temperature and pressure are not dissimilar to Earth's surface.

Were it not for the howling winds, sulphuric acid rain and the long drop it could be quite cosy! Mars has a much thinner atmosphere, and is much colder. It has clouds of water ice, but also carbon dioxide (CO_2) ice.

Titan is Saturn's largest moon, and the only other object besides Earth with liquid on its surface. Its atmosphere is thicker than Earth's, but is far too cold for liquid water to exist anywhere. Instead, Titan has a hydrocarbon cycle, with clouds, lakes, rivers and even rain made from methane and ethane.

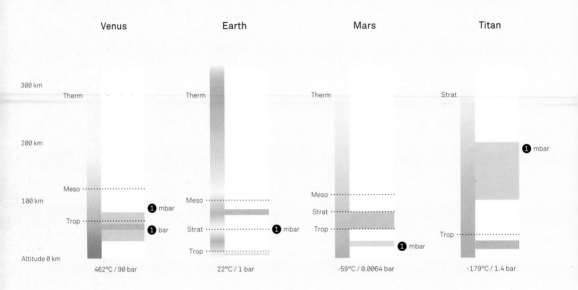

400°C	● Suphuric acid haze
0°C	● Sulphur clouds
-200°C	

● Water ice clouds
● Water

● CO_2 ice
● Water ice

● Tholin haze layer
● Methane clouds

Thermosphere, Mesosphere, Troposphere and Stratosphere

With no known surfaces on the outer planets, altitude is normally given relative to the point at which atmospheric pressure is one bar, the same as on Earth's surface.

Jupiter's striped appearance is thanks to white clouds of ammonia ice above brownish hues of ammonia hydrosulphide, below which are thought to lie water ice clouds. In 1995 the Galileo spacecraft dropped a probe into Jupiter's atmosphere which sampled the composition down to depths of over 100 kilometres and measured wind speeds over 500 kilometres per hour.

Saturn's atmosphere is similar to Jupiter's, but due to the lower gravity is more expanded in altitude. A haze layer of hydrocarbons obscures the lower structures from view, giving it a much plainer appearance than Jupiter.

Uranus and Neptune have thick layers of methane which give them their blue hue. Neptune has the strongest winds ever measured in the Solar System, at more than 1,000 kilometres per hour.

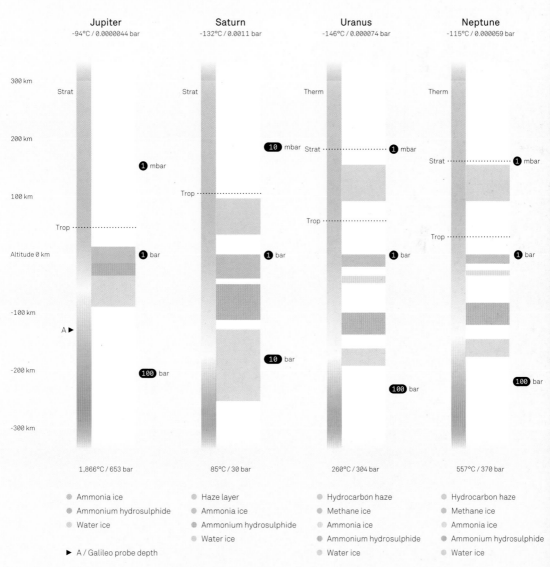

Jupiter
-94°C / 0.0000044 bar

1,866°C / 653 bar

- Ammonia ice
- Ammonium hydrosulphide
- Water ice

▶ A / Galileo probe depth

Saturn
-132°C / 0.0011 bar

85°C / 30 bar

- Haze layer
- Ammonia ice
- Ammonium hydrosulphide
- Water ice

Uranus
-146°C / 0.000074 bar

260°C / 304 bar

- Hydrocarbon haze
- Methane ice
- Ammonia ice
- Ammonium hydrosulphide
- Water ice

Neptune
-115°C / 0.000059 bar

557°C / 370 bar

- Hydrocarbon haze
- Methane ice
- Ammonia ice
- Ammonium hydrosulphide
- Water ice

Lords of the rings

Saturn is famous for its marvellous ring system, but it is not alone. The other outer planets all have their own ring systems, albeit not as bright or as expansive as Saturn's. Their origin is unknown, thought to be either the result of moons that were ripped apart, or possibly material that was never able to form a moon in the first place.

Neptune

Neptune's rings were first discovered in the 1980s, and confirmed by the *Voyager 2* spacecraft in 1989. They are named after astronomers who helped discover Neptune. Like the other planets' ring systems, they are associated with the smaller moons.

Uranus

Uranus' rings were first discovered in 1977, when they were seen to ever-so-slightly block the light of a background star. Rather than names, the rings have numbers and Greek letters.

Jupiter

Jupiter's rings are very faint and weren't discovered until the *Voyager 1* spacecraft passed by in 1979. They are the result of the bombardment of the inner satellites by small meteorites.

Saturn

The main rings of Saturn are labelled with letters in the order of discovery, while the gaps and divisions between them are named after notable people in the history of Saturn observations. Despite spanning hundreds of thousands of kilometres, Saturn's rings are incredibly thin, with the main rings just tens of metres thick.

Observations by the Voyager and Cassini spacecraft showed that the divisions are caused by Saturn's smaller moons. The outer-most 'E Ring' is thought to be created by particles spewing from the south pole of Enceladus.

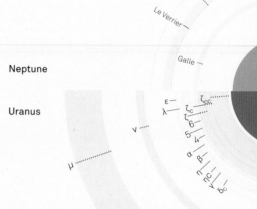

Neptune

Uranus

······ Ring width

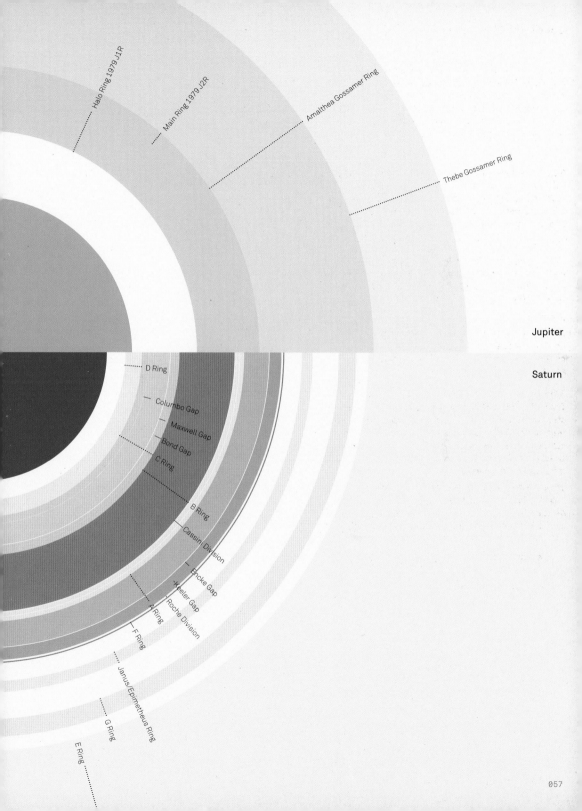

Halo Ring 1979 J1R

Main Ring 1979 J2R

Amalthea Gossamer Ring

Thebe Gossamer Ring

Jupiter

Saturn

D Ring

Columbo Gap

Maxwell Gap

Bond Gap

C Ring

B Ring

Cassini Division

Encke Gap

Keeler Gap

Roche Division

A Ring

F Ring

Janus/Epimetheus Ring

G Ring

E Ring

Asteroids

Asteroids are large lumps of rock that aren't large enough to be considered planets. Italian astronomer Giuseppe Piazzi found the first one – 1 Ceres – in 1801. In the past 200 years we've found many more smaller asteroids in similar orbits between Mars and Jupiter. These form the asteroid belt. The largest asteroids, such as Ceres, have enough mass to stay spherical but smaller asteroids, with less gravity, look much more irregular.

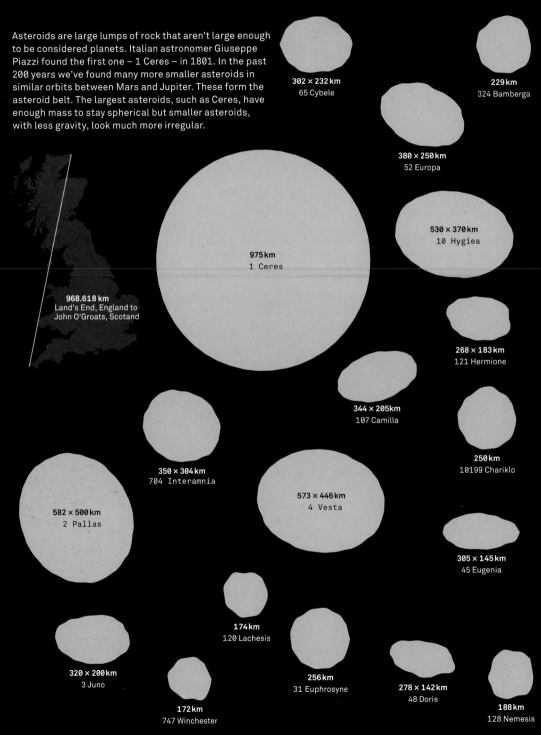

302 × 232 km
65 Cybele

229 km
324 Bamberga

380 × 250 km
52 Europa

530 × 370 km
10 Hygiea

975 km
1 Ceres

968.618 km
Land's End, England to
John O'Groats, Scotand

268 × 183 km
121 Hermione

344 × 205km
107 Camilla

250 km
10199 Chariklo

350 × 304 km
704 Interamnia

573 × 446 km
4 Vesta

582 × 500 km
2 Pallas

305 × 145 km
45 Eugenia

174 km
120 Lachesis

320 × 200 km
3 Juno

256 km
31 Euphrosyne

278 × 142 km
48 Doris

172 km
747 Winchester

188 km
128 Nemesis

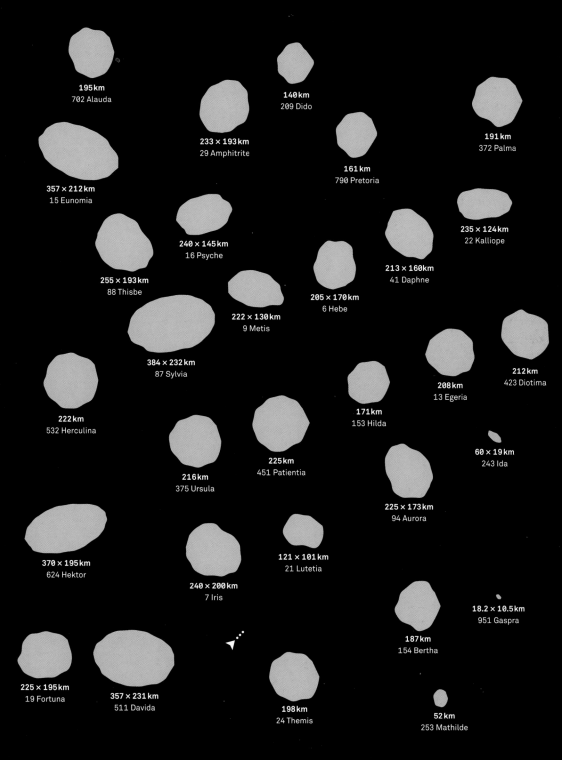

195 km
702 Alauda

140 km
209 Dido

233 × 193 km
29 Amphitrite

191 km
372 Palma

161 km
790 Pretoria

357 × 212 km
15 Eunomia

235 × 124 km
22 Kalliope

240 × 145 km
16 Psyche

255 × 193 km
88 Thisbe

213 × 160 km
41 Daphne

205 × 170 km
6 Hebe

222 × 130 km
9 Metis

384 × 232 km
87 Sylvia

212 km
423 Diotima

208 km
13 Egeria

222 km
532 Herculina

171 km
153 Hilda

60 × 19 km
243 Ida

225 km
451 Patientia

216 km
375 Ursula

225 × 173 km
94 Aurora

370 × 195 km
624 Hektor

121 × 101 km
21 Lutetia

240 × 200 km
7 Iris

18.2 × 10.5 km
951 Gaspra

187 km
154 Bertha

225 × 195 km
19 Fortuna

357 × 231 km
511 Davida

198 km
24 Themis

52 km
253 Mathilde

Locations of asteroids

The vast majority of asteroids lie in the asteroid belt
between the orbits of Mars and Jupiter, but not all
asteroids stay there, and some stray into the inner
Solar System. Those that get close to Earth's orbit,
sometimes even crossing it, are called Near-Earth
Asteroids. It is these that pose the greatest danger
to us, as one day they could hit the Earth.

Jupiter's gravitational pull has a strong influence on
the orbits of asteroids, and they don't like being in
sync with the giant planet. There are very few with
orbits longer than half that of Jupiter (a 2:1 resonance),
or shorter than a quarter of that of Jupiter (a 4:1
resonance). Those within the 4:1 resonance are
in the 'Hungaria' group.

Asteroids beyond the main belt can be trapped in
gravitational sweet spots called 'Trojan points', so
called because these asteroids are named after
participants in the Greek-Trojan war. These asteroids
stay about 60 degrees ahead or behind Jupiter in
its orbit. The Hilda family of asteroids orbit the Sun
three times for every two orbits that Jupiter makes
(a 3:2 resonance).

1 ● ● ● 90 / Main belt asteroids
1 ● ● ● 09 / Trojans & Greeks
1 ● ● ● 04 / Hildas
1 ● ● ● 03 / Near Earth Asteroids

2:1   5:2

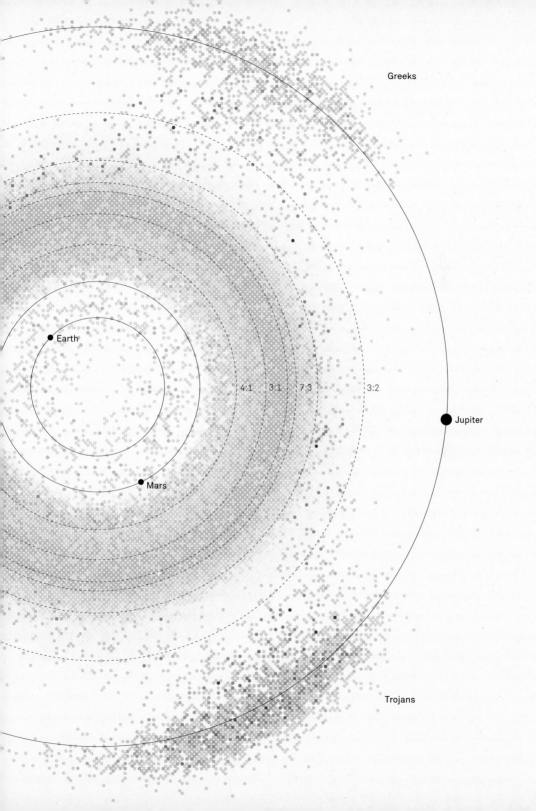

Greeks

Earth

4:1   3:1   7:3   3:2

Jupiter

Mars

Trojans

Names of asteroids

Themes have been established for naming the features, moons and rings of the major bodies in the Solar System. However, if you discover an asteroid you have the option to suggest any name you like for it. The main rule is that the name must be unique. Of 18,977 named asteroids, catalogued by the Minor Planet Center, 13,290 have citations describing who or what the asteroid is named for.

The names come from all over the world and cover a range of sources. As might be expected, many are named after scientists and astronomers (or their friends and families). There are also the names of mountains, villages, mythological characters, famous authors, and even the cast of Monty Python. Some asteroid names fall into multiple categories.

● Entertainment

13070 Seanconnery / Sean Connery (1930) British actor known from the James Bond films.

13681 Monty Python / Monty Python's Flying Circus.

246247 Sheldoncooper / Character on the TV series 'The Big Bang Theory'.

● Sport & Leisure

230975 Rogerfederer / Roger Federer (1981) Swiss tennis player.

6481 Tenzing / Tenzing Norgay (1914-1986) Nepalese climber who made the first successful ascent of Everest.

20043 Ellenmacarthur / Ellen Macarthur (1976) British solo yachtswoman who sailed around the globe.

● Friends & Family

Of asteroids named after people known to the discoverer.

19% are sons and daughters.

18% are wives and husbands.

16% are friends.

16% are parents.

5% are grandchildren.

● Geography

1718 Namibia / The African country.

10958 Mont Blanc / The highest mountain in Europe.

19620 Auckland / The largest city in New Zealand.

● Art & Literature

4444 Escher / Maurits C. Escher (1898-1972) Dutch graphic artist.

10185 Gaudi / Antoni Gaudi (1852-1926) Spanish architect.

39427 Charlottebronte / Charlotte Bronte (1816-1855) British novelist and poet.

The birth years of people with asteroids named after them.

There was a sudden increase from 1984 due to US science fairs in the 2000s naming asteroids after the winners.

Science & Nature

1991 Darwin / Charles Darwin
(1809 – 1882) English naturalist.

7672 Hawking / Stephen Hawking (1942)
Theoretical physicist famous for his work
on black holes.

25275 Jocelynbell / Jocelyn Bell (1943)
British astrophysicist who discovered pulsars.

91006 Fleming / Alexander Fleming
(1881-1955) British biologist & pharmacologist
who discovered penicillin.

Collision imminent

We discover tens of thousands of asteroids every year, most of which are very small and stay far from Earth. But every now and then we get a wake-up call, and find a large asteroid that is due to get a bit too close for comfort.

In 2029, the asteroid Apophis will pass within about 40,000 kilometres of Earth. That's really not very far astronomically speaking – only around one-tenth of the distance to the Moon. If it got just a little closer there would be huge devastation.

But then there are those asteroids we don't see coming. In 1908 an asteroid impacted in Tunguska, Russia, flattening trees over a huge area. We're on the lookout for these objects now, and though we have discovered hundreds in recent years, we still don't catch them all.

In February 2013 the world was taken by surprise when an asteroid 10 – 20 metres across entered the atmosphere with no warning and exploded over Chelyabinsk, Russia, injuring over 1,000 people. How many more are we missing?

Years 1900 – 2100

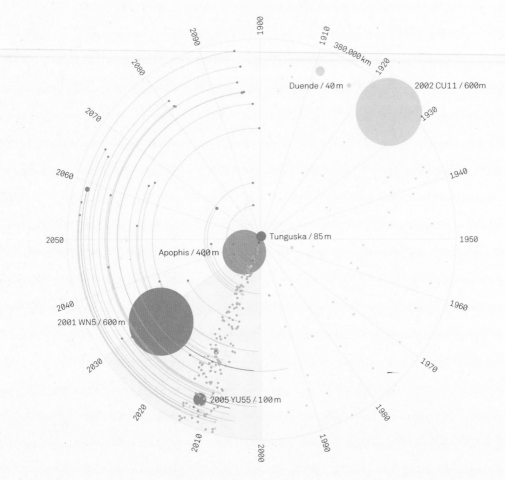

500 m / 100 m / 50 m / <20 m

- ● Impacted
- ● Flyby with at least 1 month warning
- — Warning time
- ● Flyby with little or no warning
- ● Discovered after flyby

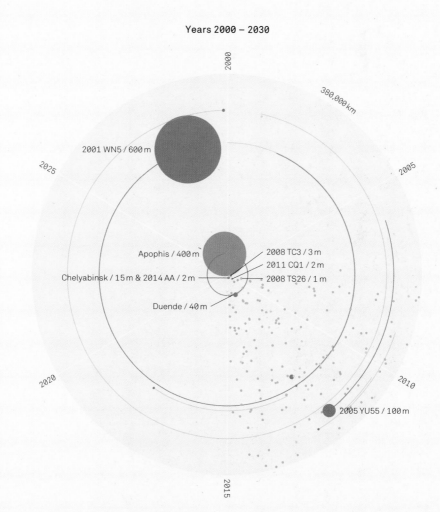

Years 2000 – 2030

2000

380,000 km

2025

2005

2001 WN5 / 600 m

Apophis / 400 m

2008 TC3 / 3 m

2011 CQ1 / 2 m

Chelyabinsk / 15 m & 2014 AA / 2 m

2008 TS26 / 1 m

Duende / 40 m

2020

2010

2005 YU55 / 100 m

2015

Classifications of meteorites

Now and again small pieces of debris are large enough to make it down to Earth without burning up or exploding in the atmosphere. A small fraction are observed as they fall from the sky, though most are found lying on the ground, most commonly on top of the Antarctic ice sheet. Some are made of rock, others iron, and some both. Studying the chemical make-up of these fragments can sometimes even tell us which object they come from.

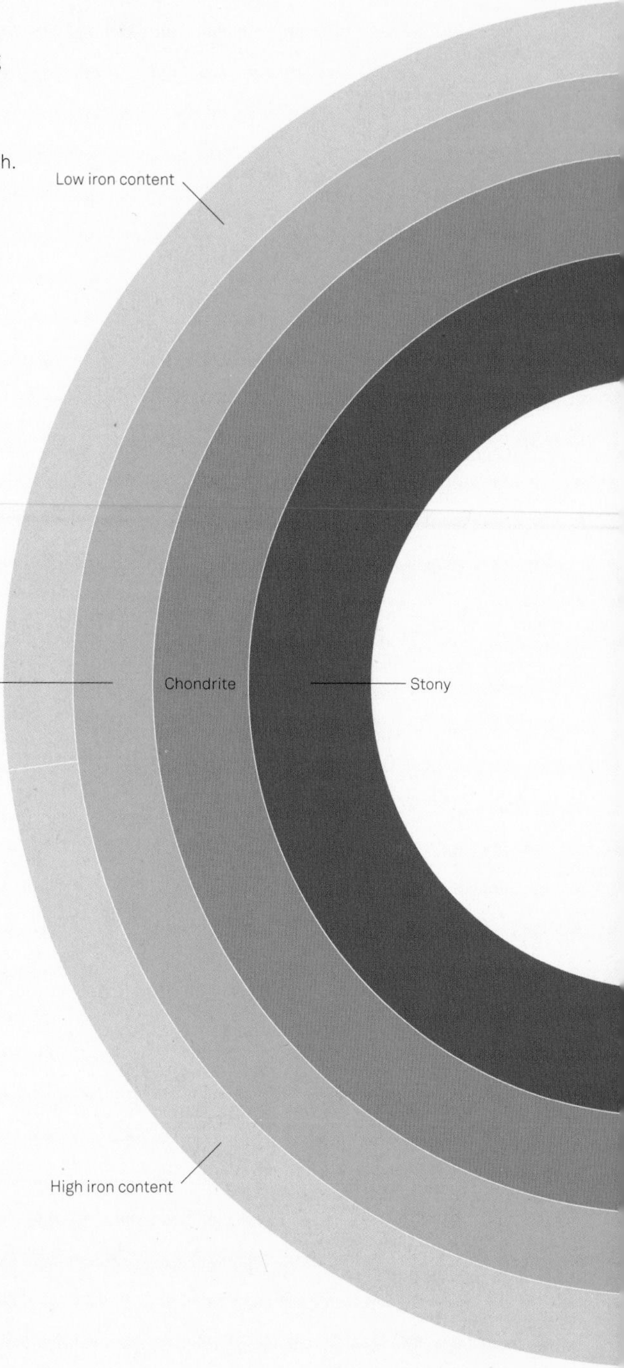

Low iron content

Stony meteorites

The vast majority of meteorites are made predominantly of rock. The materials and minerals are similar to those in the Earth's crust, though normally with small amounts of metals such as iron.

Ordinary chondrite — — Chondrite — — Stony

Chondrites

Chondrites are stony meteorites that contain small spherical inclusions of specific minerals. They have been heated slightly, meaning that they have not been altered since the birth of the Solar System. The most primitive of these are carbonaceous chondrites.

High iron content

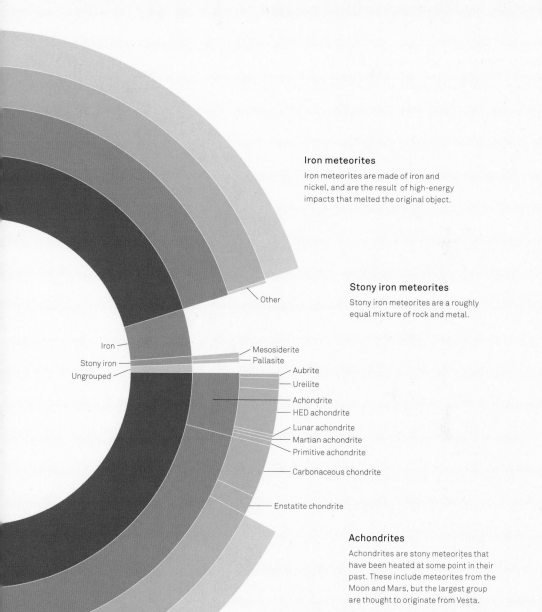

Iron meteorites

Iron meteorites are made of iron and nickel, and are the result of high-energy impacts that melted the original object.

Stony iron meteorites

Stony iron meteorites are a roughly equal mixture of rock and metal.

Other

Iron

Stony iron

Ungrouped

Mesosiderite

Pallasite

Aubrite

Ureilite

Achondrite

HED achondrite

Lunar achondrite

Martian achondrite

Primitive achondrite

Carbonaceous chondrite

Enstatite chondrite

Achondrites

Achondrites are stony meteorites that have been heated at some point in their past. These include meteorites from the Moon and Mars, but the largest group are thought to originate from Vesta.

A comet

Thought to be omens by the ancients, we now know that comets are small icy and rocky bodies that orbit the Sun on elongated orbits. Although famous for their tails, they spend most of their time in the outer Solar System where it is cold and their tails are inactive. As they fall inwards towards the Sun the temperature rises. Their outer layers start to sublimate away from their central nuclei and form expansive tails. Comets actually have two tails; a dust tail and an ion tail.

As of 2015, six comets have been visited by spacecraft. In 2014 ESA's Rosetta spacecraft visited comet *67P/ Churyumov-Gerasimenko*. It dropped the Philae lander which completed the first ever soft landing on a comet.

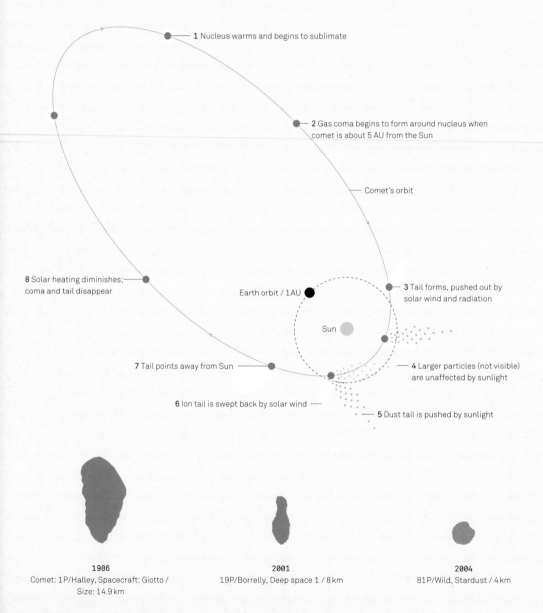

1 Nucleus warms and begins to sublimate

2 Gas coma begins to form around nucleus when comet is about 5 AU from the Sun

Comet's orbit

8 Solar heating diminishes; coma and tail disappear

Earth orbit / 1AU

3 Tail forms, pushed out by solar wind and radiation

Sun

7 Tail points away from Sun

4 Larger particles (not visible) are unaffected by sunlight

6 Ion tail is swept back by solar wind

5 Dust tail is pushed by sunlight

1986
Comet: 1P/Halley, Spacecraft: Giotto / Size: 14.9km

2001
19P/Borrelly, Deep space 1 / 8km

2004
81P/Wild, Stardust / 4km

—— Tail approximately 100 million km long

Coma approximately ———
1 million km across

——— Hydrogen envelope approximately
10 million km across

Icy nucleus 1-10 km in diameter

2005
9P/Tempel, Deep impact / 7.6 km

2010
103P/Hartley, Epoxi / 1.6 km

2014
67P/Churyumov-Gerasimenko, Rosetta / 4.3 km

Comets

We only generally see comets when they come into the inner Solar System. But they originate much further out. Comets are split into several main groups depending on their orbits.

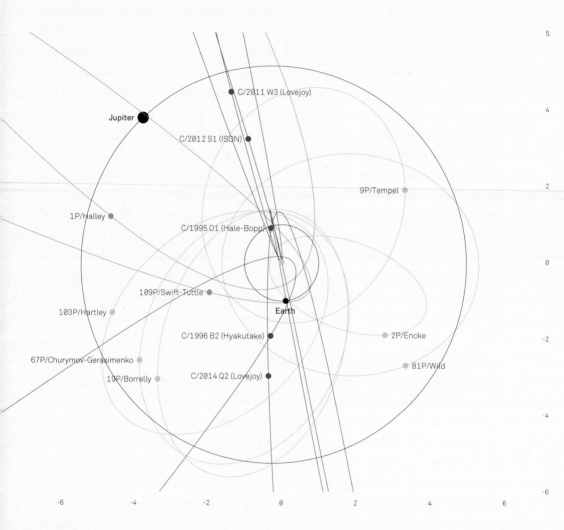

● Short period comets

The short period comets have orbital periods less than 200 years, with most staying within the Kuiper Belt. Some, such as Halley's Comet (*1P/Halley*) come in at relatively high angles, and are though to have originated in the Oort cloud.

● Jupiter family comets

The Jupiter family comets are short-period comets with orbits less than 20 years long, and which tend to orbit in the same plane as the planets. They are thought to have formed in the Kuiper Belt, but been sent inwards after coming too close to the giant planets. They are now on orbits that take them about as far out as Jupiter's orbit.

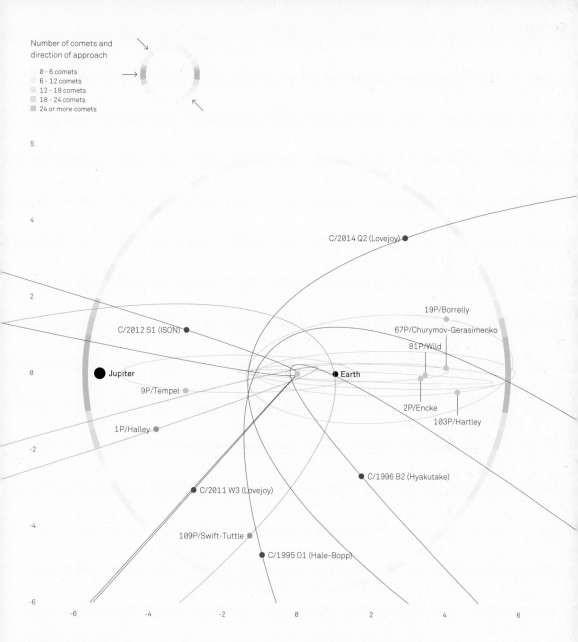

Number of comets and direction of approach

- 0 - 6 comets
- 6 - 12 comets
- 12 - 18 comets
- 18 - 24 comets
- 24 or more comets

C/2014 Q2 (Lovejoy)

19P/Borrelly
67P/Churymov-Gerasimenko
81P/Wild
C/2012 S1 (ISON)

Jupiter   Earth
9P/Tempel
2P/Encke
103P/Hartley

1P/Halley

C/1996 B2 (Hyakutake)

C/2011 W3 (Lovejoy)

109P/Swift-Tuttle
C/1995 O1 (Hale-Bopp)

● Long period comets

The long period comets have orbits over 200 years long, typically taking thousands of years. These originated in the Oort Cloud, but have been scattered in towards the inner Solar System. 'Sungrazing comets' are those which get very close to the Sun, such as *C/2011 W3 (Lovejoy)*. Most are thought to be fragments of a larger comet that broke up thousands of years ago when it got too close to the Sun.

● Non-periodic comets

The non-periodic comets are on such long orbits that it's not clear whether they will ever return to the inner Solar System or, if they do, not for millions of years. They may escape the Sun's gravity entirely. Comet *C/2012 S1 (ISON)* made its first visit to the inner Solar System in late 2013. It was a sungrazing comet that got so close to the Sun that it was ripped apart, leaving behind a cloud of gas and dust.

Comet hunters

Comets – from the Greek 'kometes' meaning 'long–haired' – have been known since antiquity. For centuries it had been argued that they were anything from refracted light, to atmospheric vapours, to chains of stars. We now know that they orbit the Sun and many are periodic in nature.

The invention of telescopes spurred a spate of comet discoveries. One of the earliest comet discoverers was Caroline Herschel who found a remarkable eight comets. Jean-Louis Pons set a long-held record for the most comets found by one person with 37. This was nearly equalled in the 1980/90s by Caroline and Eugene Shoemaker who co-discovered the famous *Shoemaker-Levy 9* which hit Jupiter in 1994. In the 21st Century, Robert H. McNaught has smashed the record with 82 comets to his name.

The 21st Century has seen the rise of ground and space-based projects which have often found comets as a by-product of their main observations. Most notable are discoveries on a mission by the ESA/NASA SOHO satellite designed to monitor the Sun: over 2,800 comets have been found, thanks to amateur and professional comet hunters combing images from the spacecraft. Perhaps the most prolific amateur is British astronomer Mike Oates who has found 144 comets this way.

Method of discovery and total number to date
- Human
- Robotic survey
- Spacecraft

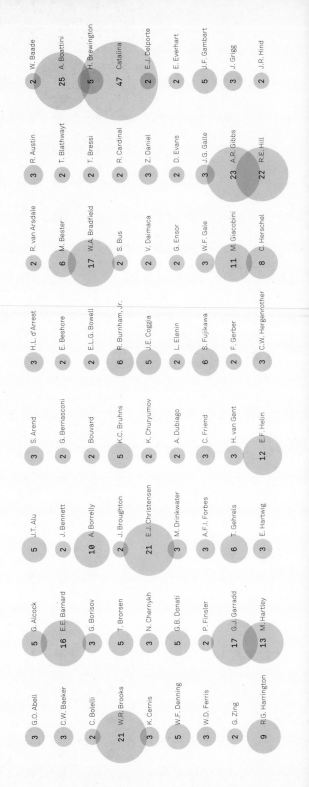

6 IRAS | 6 E. Klinkerfues | 2 S. Kozik | 17 Lemmon | 10 D. Machholz | 12 C. Messier | 2 S. Murakami | 2 Palomar | 2 E. Peterson | 5 B.P. Roman | 14 STEREO | 2 H.E. Schuster | 32 E. Shoemaker | 14 Swift | 7 D. du Toit | 4 Y. Väisälä | 6 F.L. Whipple | 2 T. Yanaka

10 K. Ikeya | 2 D. Klinkenberg | 9 R. Kowalski | 3 K. Lawrence | 6 T. Lovejoy | 5 J.E. Mellish | 15 J. Mueller | 5 L. Pajdušáková | 2 C. Peters | 3 L. Respighi | 19 SOLWIND | 3 A. Schaumasse | 32 C.S. Shoemaker | 2 M. Suzuki | 2 J. Tilbrook | 2 S. Utsunomiya | 3 R.M. West | 3 M. Wolf

2 M.L. Humason | 2 T. Kiuchi | 6 C.T. Kowal | 6 S. Larson | 5 M. Lovas | 4 R. Meier | 12 A. Mrkos | 4 L. Oterma | 9 C.D. Perrine | 2,842 SOHO | 2 P. Shajn | 3 T.B. Spahr | 2 J. Thiele | 6 H.P. Tuttle | 2 F.G. Watson | 5 C.A. Wirtanen

2 D. Hughes | 2 M. Jäger | 3 K. Korlevic | 3 C.I. Lagerkvist | 20 LONEOS | 7 P. Méchain | 2 H. Mori | 3 H.W.M. Olbers | 2 F. Pereyra | 2 K.W. Reinmuth | 20 SMM | 2 J.M. Schaeberle | 9 T. Seki | 29 Spacewatch | 6 Tenagra | 2 R. Tucker | 4 A.A. Wachmann | 10 F.A.T. Winnecke

12 M. Honda | 2 C. Juels | 2 A. Kopff | 222 LINEAR | 4 J. Montani | 7 G. Neujmin | 12 L. Pelter | 6 W. Reid | 2 K. Rümker | 4 Y. Sato | 10 J.V. Scotti | 2 C.D. Slaughter | 12 W. Tempel | 2 A.F. Tubbiolo | 18 WISE | 7 A.G. Wilson

5 P.R. Holvorcem | 3 A.F.A.L. Jones | 2 N. Kojima | 8 La Sagra | 2 E. Liais | 82 R.H. McNaught | 2 J. Montaigne | 2 A. Nakamura | 2 J. Paraskevopoulos | 2 M. Read | 13 K.S. Russell | 2 A. Sandage | 4 K.G. Schweizer | 5 J.F. Skjellerup | 5 K. Takamizawa | 3 Tsuchinshan | 6 F. de Vico | 4 A. Wilk

6 H.E. Holt | 4 E. Johnson | 5 L. Kohoutek | 2 Y. Kushida | 2 W. Li | 2 R.S. McMillan | 2 M. Mitchell | 3 NEOWISE | 55 PANSTARRS | 2 F. Quénisset | 3 M. Rudenko | 2 Y. Saigusa | 4 F.K.A. Schwassmann | 16 B.A. Skiff | 3 A. Tago | 2 K. Tritton | 4 M.E. Van Ness | 7 P. Wild

2 E. Holmes | 3 C. Jackson | 2 T. Kobayashi | 2 L. Kresák | 22 D.H. Levy | 2 A. Maury | 5 J.H. Metcalf | 54 NEAT | 37 J.L. Pons | 2 D. Ross | 10 SWAN | 3 M. Schwartz | 13 Siding Spring | 3 V. Tabur | 3 C. Torres | 3 G. Van Biesbroeck | 2 G.L. White

Kuiper belt

In 1992 astronomers discovered a small object
out beyond the orbit of Neptune. Called 1992 QB1,
it is now one of around 1,000 known objects that
join Pluto out in the Kuiper Belt. The discovery of
Eris in 2005 led to the realisation that Pluto is not
the only large object in the outer Solar System.
Most move in ellipses rather than circles, changing
their distance from the Sun as they orbit.

Pluto / plan

Objects in the Kuiper Belt often have orbits tilted
relative to the eight main planets, so they move up
and down as well as towards and away from the Sun.

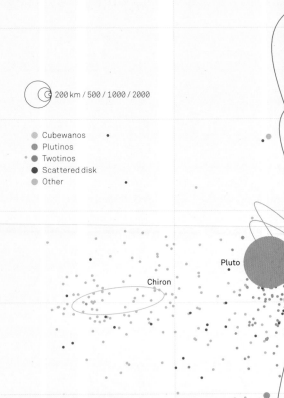

200 km / 500 / 1000 / 2000

- Cubewanos
- Plutinos
- Twotinos
- Scattered disk
- Other

Pluto

Chiron

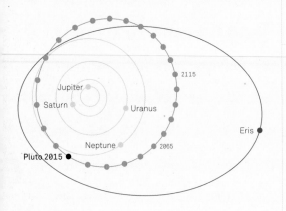

Jupiter
Saturn
Uranus
Neptune
2115
2065
Pluto 2015
Eris

Pluto / elevation

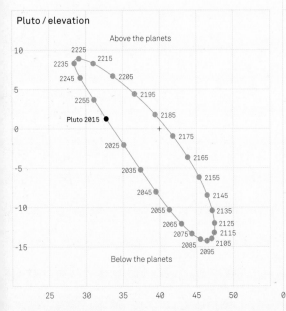

Above the planets

10
5
0
-5
-10
-15

Below the planets

2225 2215
2235
2245 2205
2255
2195
2185
Pluto 2015
2025 2175
2165
2035 2155
2045 2145
2055 2135
2065 2125
2075 2115
2085 2105
2095

25   30   35   40   45   50

Most objects in the Kuiper Belt orbit just
beyond Neptune and don't cross its orbit
(unlike Pluto). These are called 'Cubewanos'
after the first discovered object, *1992 QB1*.

There are groups of objects that have orbits
that resonate with Neptune's. Like Pluto,
'Plutinos' orbit the Sun twice for every three
times Neptune orbits. 'Twotinos' orbit once
for every two Neptune orbits.

Some bodies have very wild orbits, thought
to be as a result of the gravitational pull of the
outer planets, particularly Jupiter and Neptune.
These make up the Scattered Disk.

0 Distance from Sun (x Earth-Sun distance)   20

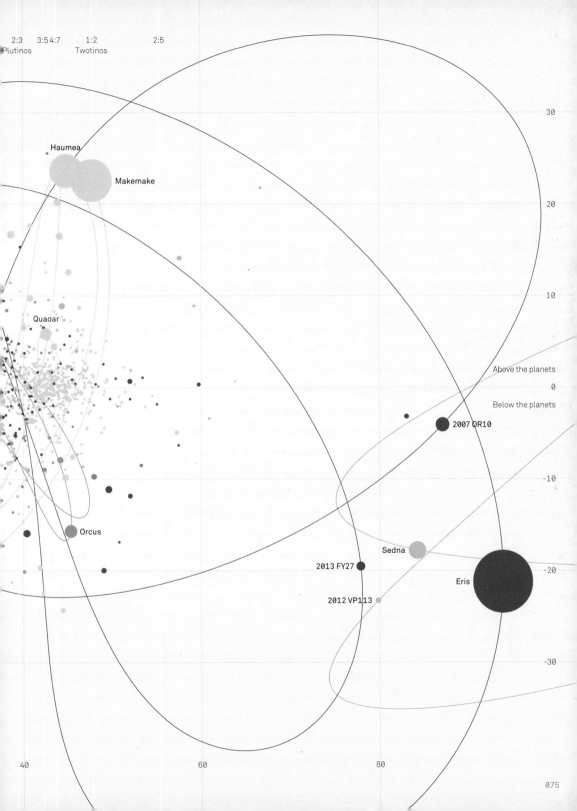

30

20

10

Haumea

Makemake

Quaoar

Above the planets

0

Below the planets

2007 OR10

-10

Orcus

Sedna

2013 FY27

-20

Eris

2012 VP113

-30

40   60   80

The high jump

If you can jump half a metre high on the Earth, how high
could you jump on the Moon, Jupiter, or an asteroid?
That depends on the mass and size of the object. If it's
too small and light you'll never come back down.

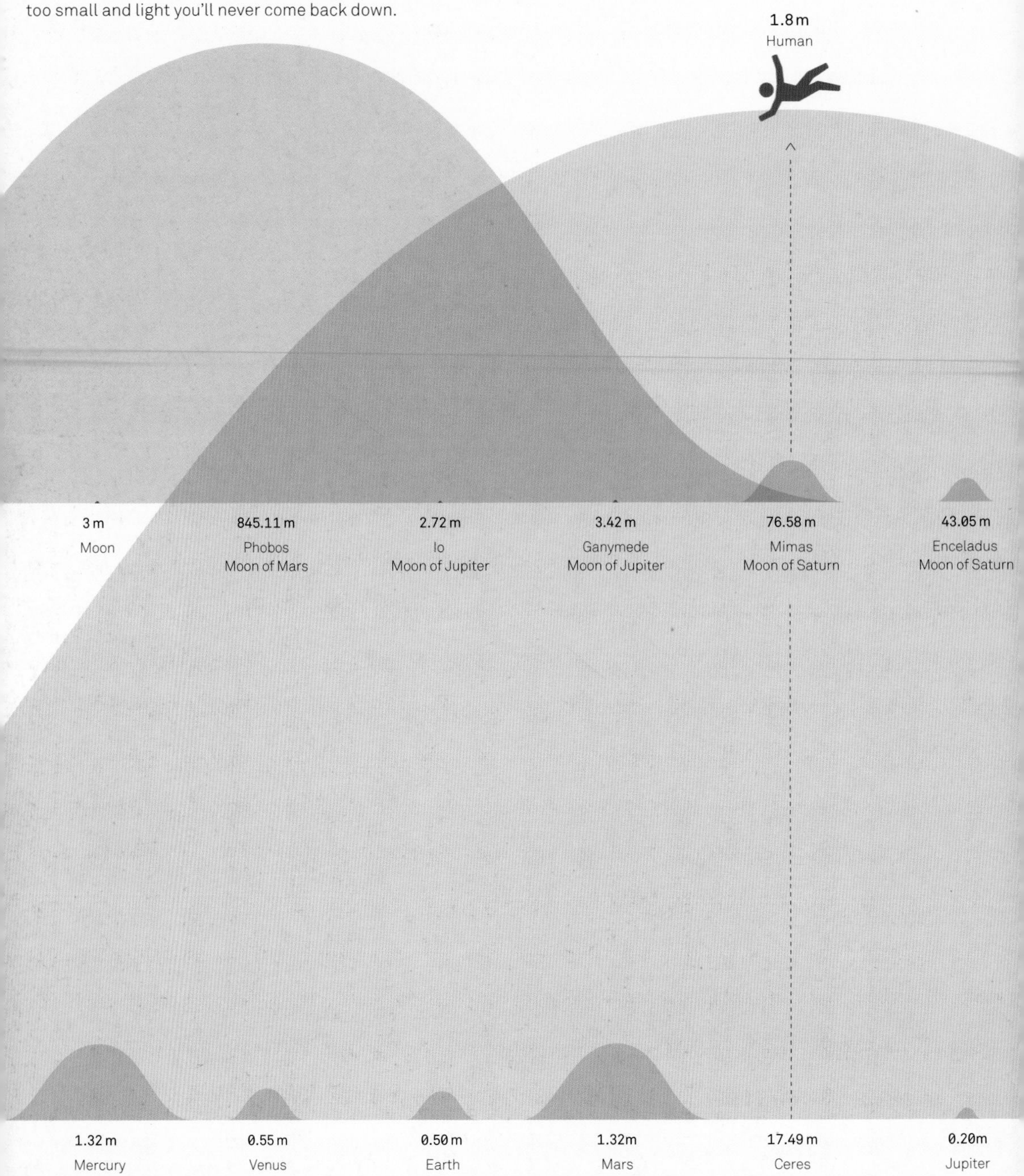

1.8 m
Human

3 m
Moon

845.11 m
Phobos
Moon of Mars

2.72 m
Io
Moon of Jupiter

3.42 m
Ganymede
Moon of Jupiter

76.58 m
Mimas
Moon of Saturn

43.05 m
Enceladus
Moon of Saturn

1.32 m
Mercury

0.55 m
Venus

0.50 m
Earth

1.32m
Mars

17.49 m
Ceres

0.20m
Jupiter

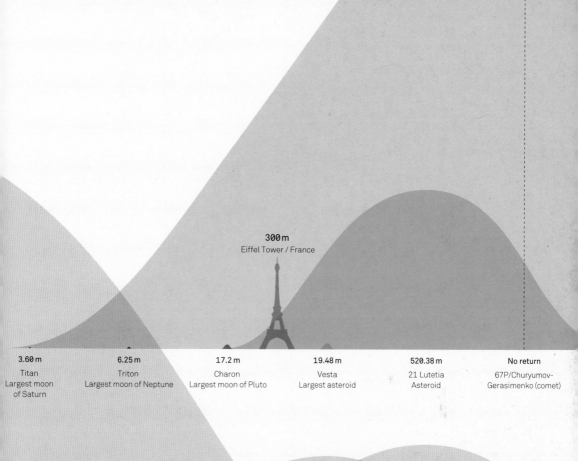

300 m
Eiffel Tower / France

3.60 m	6.25 m	17.2 m	19.48 m	520.38 m	No return
Titan	Triton	Charon	Vesta	21 Lutetia	67P/Churyumov-
Largest moon	Largest moon of Neptune	Largest moon of Pluto	Largest asteroid	Asteroid	Gerasimenko (comet)
of Saturn					

0.47 m	0.55 m	0.44 m	7.42 m	7.78 m	5.92 m
Saturn	Uranus	Neptune	Pluto	Haumea	Eris

Timeline of the Solar System

We know the age of the Solar System fairly precisely thanks to evidence from meteorites, which contain the first material to solidify. A lot has happened since then...

⇄ Ice age ⇄ Astronomy ⇄ Geology ⇄ Life ☠ Extinction event

A -4,568 M yr / Formation of asteroids and comets
B -4,400 M yr / Formation of Saturn's rings
   -4,400 M yr / Oldest minerals on Earth
C -4,100 M yr / Possible primordial life
D -4,000 M yr / Oldest rocks on Earth
E -3,600 M yr / First simple single-celled life & microfossils
F -2,300 M yr / Appearance of oxygen in Earth's atmosphere
G -2,100 M yr / First photosynthesis

-4,568 – 4,564 M yr (million years) / Formation of the giant planets
-4,568 – 4,558 M yr / Terrestrial planets formed
-4,563 – 4,553 M yr / Gas and dust disk depleted
-4,568 – 4,000 M yr / Hadean eon
-4,508 – 4,478 M yr / Formation of the Moon
A
4,500 M yr

-4,300 – 4,100 M yr / Formation of the large lunar basins

-4,000 – 2,500 M yr / Archaen eon
E
-3,500 M yr
-3,768 – 3,668 M yr / Uranus and Neptune switch places

-3,000 M yr

-1,500 M yr
H
-1,000 M yr

-420 – 370 M yr / First ferns trees & seed-bearing plants
I
-370 – 325 M yr / First land vertebrates
-325 – 300 M yr / First reptiles; Coal forests; Highest ever atmospheric oxygen levels
☠ 70% of species

L ☠
-200 – 66 M yr / Dinosaurs rule the world
70-75% of species
-100 M yr

-56 – 35 M yr / Seafloor algae lower atmospheric CO$_2$
-66 – 57 M yr / First large mammals & primates
☠ 75% of species
-50 M yr
M
-40 M yr

+ 50 – 60 M yr / Canadian Rockies erode away
100 M yr   T   S
Now   R   Q   P   O   N
-2.6 – 0 M yr / Current ice age
U
250 M yr
V
500 M yr
2,000 M yr

2,500 M yr
4,000 – 5,000 M yr / Andromeda Galaxy & Milky Way merge; 12% chance Sun will be ejected from 'Milkomeda'
4,000 M yr
4,500 M yr

6,000 M yr
6,500 M yr

H -1,000 M yr / First simple multi-celled fossils
I -465 M yr / First green plants and fungi
J -300 M yr / Pangea supercontinent forms
K -250 M yr / First dinosaurs, crocodiles & mammals
L -200 M yr / Pangea breaks into Gondwana & Laurasia
M -50 M yr / Beginning of formation of Himalayas
N -8 M yr / Evolutionary split from gorillas
O -4 M yr / Evolutionary split from chimpanzees

P -2.3 M yr / First hominids
Q -1.4 M yr / First appearance of Homo erectus
R -0.2 M yr / First appearance of Homo sapiens
S 50 M yr / Phobos crashes into Mars or breaks up into a ring
T 80 M yr / Big Island of Hawaii will sink beneath the ocean
U 250 M yr / Another supercontinent forms
V 600 M yr / Moon too far for total solar eclipses to occur
W 3,500 M yr / Earth's atmosphere more like Venus' is now

-4,468 – 4,068 M yr / Jupiter & Saturn fall into resonance

-4,368 – 4,268 M yr / Cluster of stars disperses

B

C

-4,068 – 3,868 M yr / Late heavy bombardment

D

-4,000 M yr

-2,800 – 2,500 M yr / Stabilisation of the Earth's tectonic plates

-2,500 – 2,100 M yr / Ice age

-2,500 – 540 M yr / Protozoic eon

G

F

-2,000 M yr

-840 – 630 M yr / Ice age

-540 – 0 M yr / Phanerozoic eon

-500 M yr

-445 – 420 M yr / First jawed fish

-460 – 420 M yr / Ice age

60-70% of species

-360 – 260 M yr / Ice age

-540 – 485 M yr / Cambrian explosion

J

K

-250 M yr / 90-96% of species

-75 M yr

-34 – 23 M yr / Rapid evolution of mammals

-30 M yr

-23 – 7 M yr / Widespread forests lower atmospheric CO_2 levels

-20 M yr

1,000 M yr

1,000 – 2,000 M yr / Increased energy output of Sun causes oceans to boil away

1,500 M yr

3,000 M yr

W

3,500 M yr

5,000 M yr

5,420 – 7,720 M yr / Sun grows into red giant and may engulf the Earth

5,500 M yr

7,500 M yr

Travel times

How long would it take to travel the Solar System? It depends how fast you go. If it were possible, a drive to the Moon would take about half a year at 100 kilometres per hour whereas it would take just over a second travelling at the speed of light.

● Planets / dwarf planets
● Stars
● Other

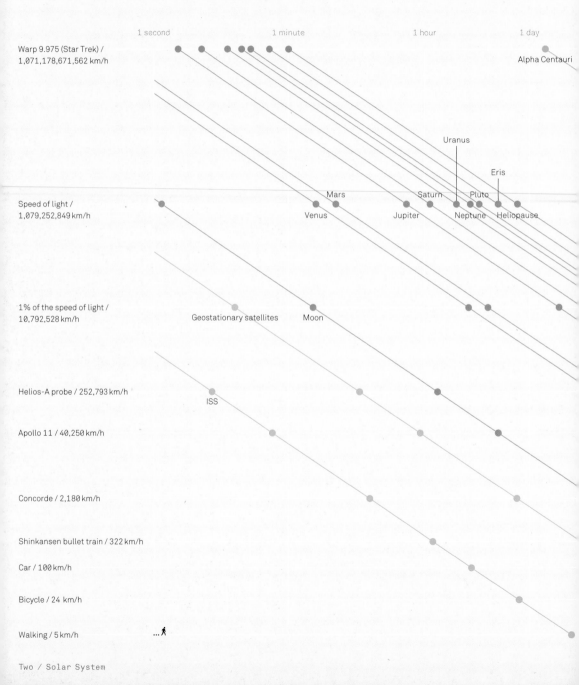

1 second   1 minute   1 hour   1 day

Warp 9.975 (Star Trek) /
1,071,178,671,562 km/h
Alpha Centauri

Uranus

Eris

Mars   Saturn   Pluto
Venus   Jupiter   Neptune   Heliopause

Speed of light /
1,079,252,849 km/h

1% of the speed of light /
10,792,528 km/h
Geostationary satellites   Moon

Helios-A probe / 252,793 km/h
ISS

Apollo 11 / 40,250 km/h

Concorde / 2,180 km/h

Shinkansen bullet train / 322 km/h

Car / 100 km/h

Bicycle / 24 km/h

Walking / 5 km/h
... 🚶

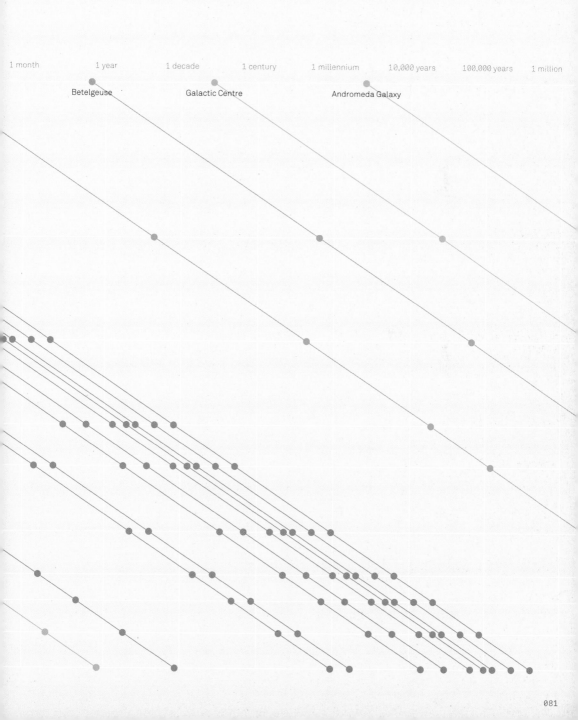

1 month   1 year   1 decade   1 century   1 millennium   10,000 years   100,000 years   1 million

Betelgeuse   Galactic Centre   Andromeda Galaxy

Three / Telescopes

Optical telescopes / size matters

There are two things that every telescope does, whatever it looks like. Firstly, it collects the light that falls through the main aperture, whether that is through a lens or a mirror. Secondly, it focuses that light onto a camera, film, eyepiece, or some other sort of detector. There are many ways of doing this, which often involve additional lenses or mirrors, and result in telescopes with a wide range of appearances, but those two principal jobs of a telescope haven't changed since the first one was used in the early 17th Century. Much of the development in these instruments over the last four centuries has revolved around making the main mirror or lens bigger.

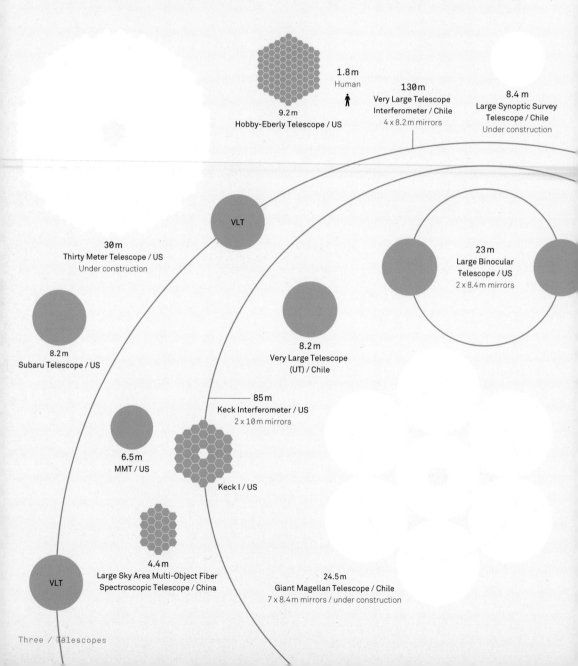

9.2 m
Hobby-Eberly Telescope / US

1.8 m
Human

130 m
Very Large Telescope
Interferometer / Chile
4 x 8.2 m mirrors

8.4 m
Large Synoptic Survey
Telescope / Chile
Under construction

30 m
Thirty Meter Telescope / US
Under construction

VLT

23 m
Large Binocular
Telescope / US
2 x 8.4 m mirrors

8.2 m
Subaru Telescope / US

8.2 m
Very Large Telescope
(UT) / Chile

85 m
Keck Interferometer / US
2 x 10 m mirrors

6.5 m
MMT / US

Keck I / US

4.4 m
Large Sky Area Multi-Object Fiber
Spectroscopic Telescope / China

VLT

24.5 m
Giant Magellan Telescope / Chile
7 x 8.4 m mirrors / under construction

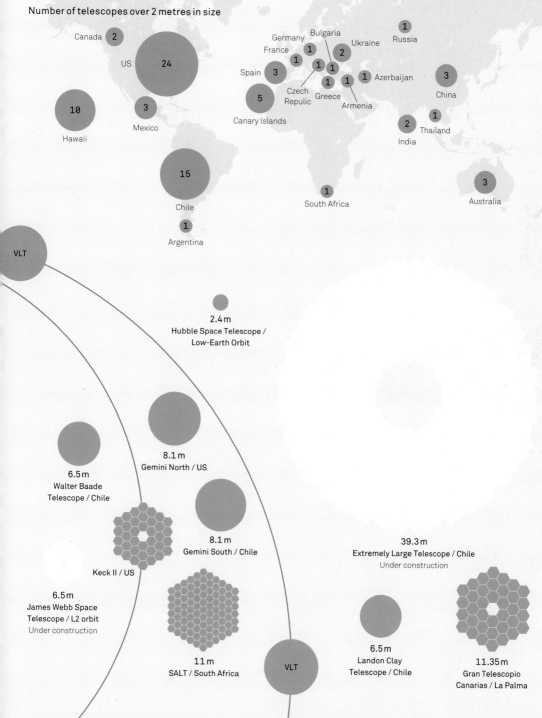

Number of telescopes over 2 metres in size

Canada **2**

US **24**

Hawaii **10**

Mexico **3**

Germany

France

Bulgaria

Ukraine **2**

Russia **1**

Spain **3**

Czech Repulic

Greece

Armenia

Azerbaijan **1**

China **3**

Canary Islands **5**

Thailand **1**

India **2**

Chile **15**

South Africa **1**

Australia **3**

Argentina **1**

VLT

2.4 m
Hubble Space Telescope /
Low-Earth Orbit

8.1 m
Gemini North / US

6.5 m
Walter Baade
Telescope / Chile

8.1 m
Gemini South / Chile

Keck II / US

6.5 m
James Webb Space
Telescope / L2 orbit
Under construction

39.3 m
Extremely Large Telescope / Chile
Under construction

11 m
SALT / South Africa

VLT

6.5 m
Landon Clay
Telescope / Chile

11.35 m
Gran Telescopio
Canarias / La Palma

Atmospheric windows

We think of the night sky as being full of stars, but our eyes show us only a tiny part of the full spectrum of light available. Using telescopes to view longer and shorter wavelengths we can see a huge range of objects and astrophysical phenomena. The Earth's atmosphere blocks much of that light, leaving wavelength windows that can be observed from the ground. To get the best view, telescopes have to be situated at the tops of mountains, flown on board aeroplanes, dangled from high-altitude balloons, or even launched into space.

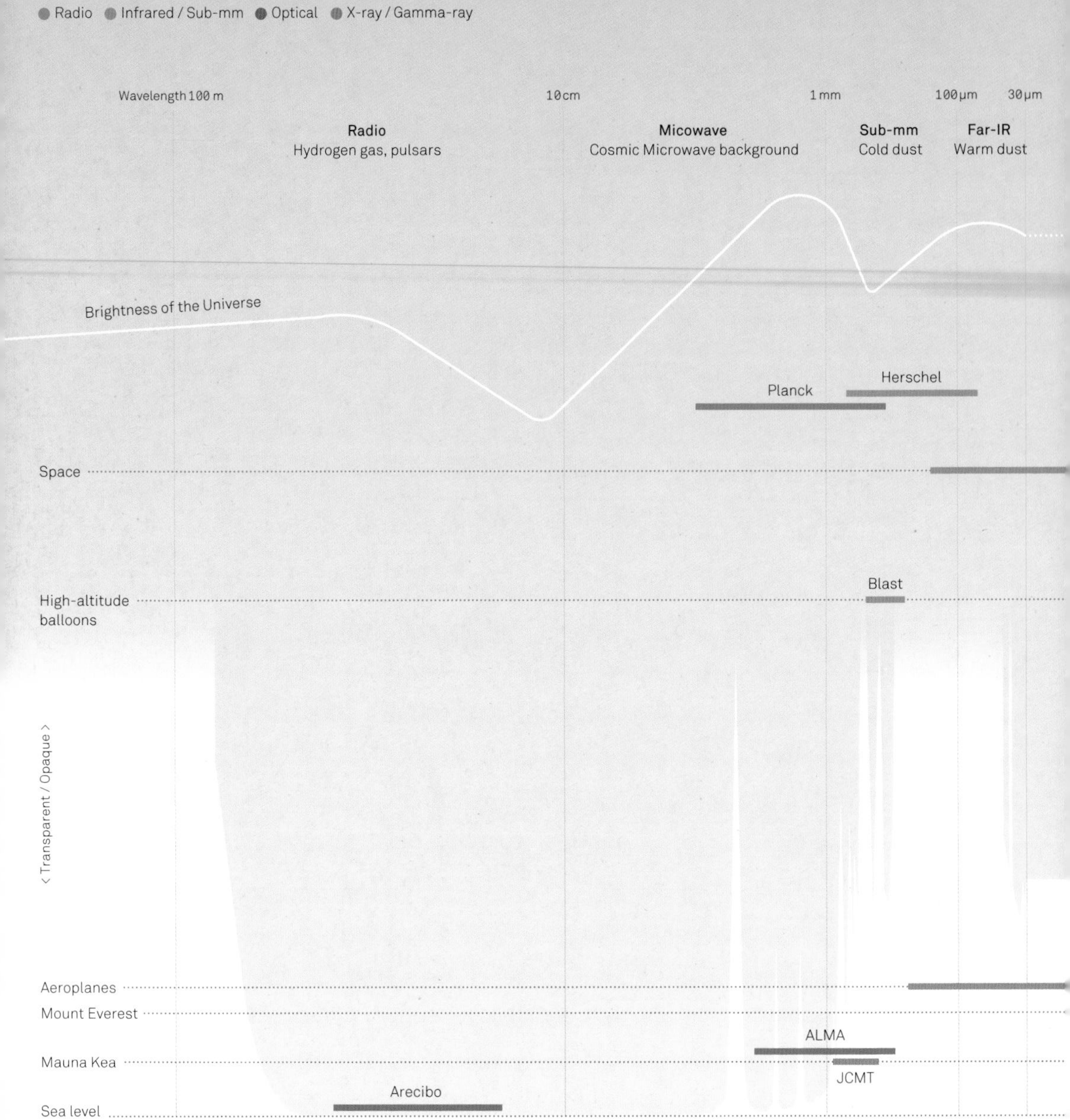

● Radio ● Infrared / Sub-mm ● Optical ● X-ray / Gamma-ray

Wavelength 100 m   10 cm   1 mm   100 μm   30 μm

Radio
Hydrogen gas, pulsars

Micowave
Cosmic Microwave background

Sub-mm
Cold dust

Far-IR
Warm dust

Brightness of the Universe

Planck

Herschel

Space

High-altitude balloons

Blast

< Transparent / Opaque >

Aeroplanes

Mount Everest

ALMA

Mauna Kea

JCMT

Sea level

Arecibo

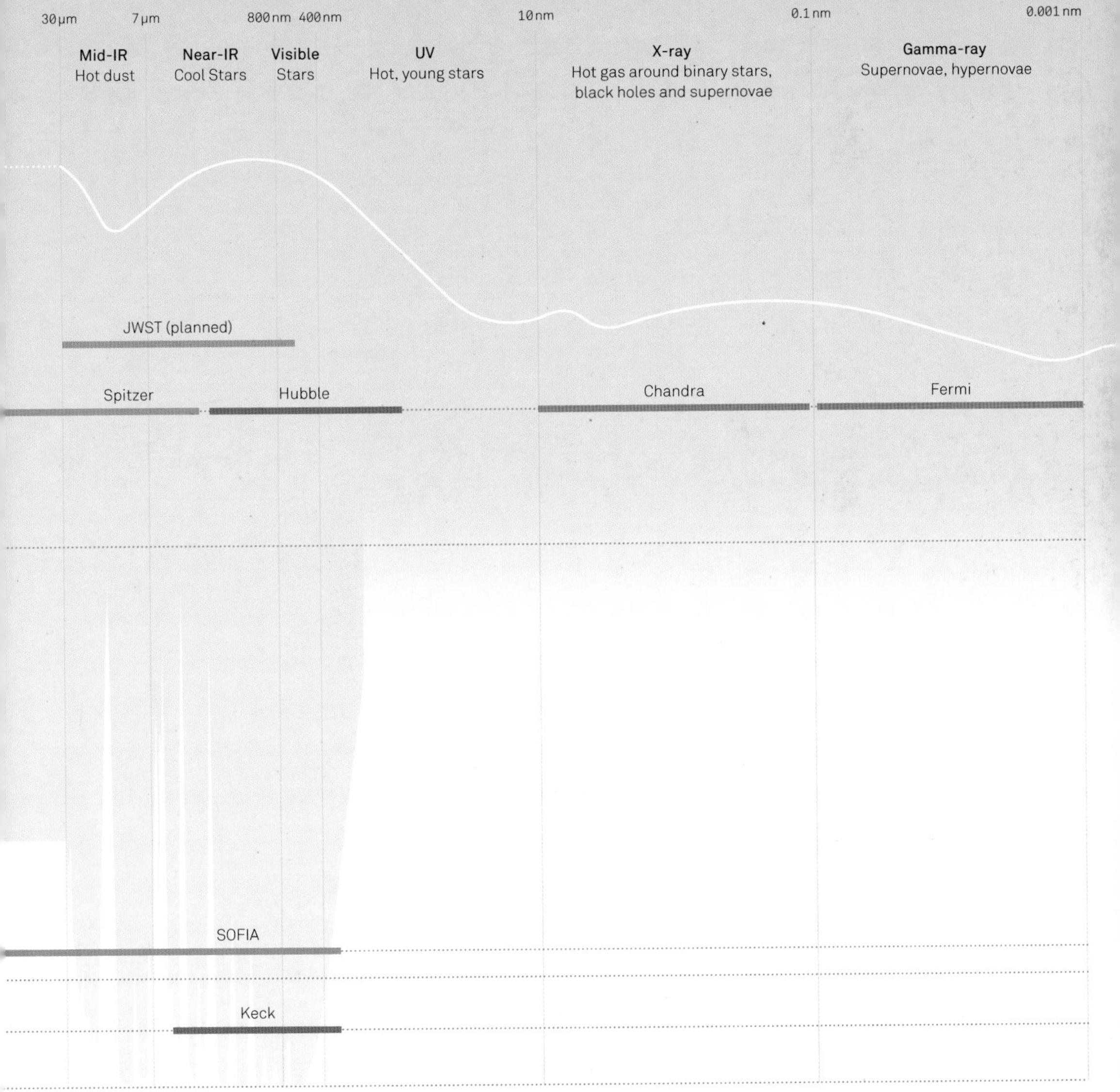

| 30 μm | 7 μm | 800 nm 400 nm | 10 nm | 0.1 nm | 0.001 nm |

Mid-IR
Hot dust

Near-IR
Cool Stars

Visible
Stars

UV
Hot, young stars

X-ray
Hot gas around binary stars,
black holes and supernovae

Gamma-ray
Supernovae, hypernovae

JWST (planned)

Spitzer

Hubble

Chandra

Fermi

SOFIA

Keck

The sky's the limit

No optical telescope, however large it is, can observe through clouds. For this reason, most large telescopes are built as high as possible – above the clouds. Even so, the biggest obstacle a telescope faces is still the Earth's own atmosphere. Most of the best locations are on top of mountains, hence the mountain-top observatories in Europe and the USA. In the second half of the 20th Century, more remote sites became heavily used such as the Canary Islands, Mauna Kea on Hawaii, and the Andes mountain range in Chile.

There is still some air even at altitudes above 4,000 metres, so many of the largest telescopes employ techniques such as adaptive optics to reach the finest possible resolutions.

At longer wavelengths, the water vapour in the atmosphere is the biggest problem, and so it is important to find a very high, dry site. The best such example is the Atacama Desert in Chile, home to a number of optical telescopes and the ALMA telescope array. Coming a close second is the South Pole, which is very dry and also, thanks to a thick layer of ice, at relatively high altitude.

Infrared/sub-mm telescope

Optical telescope

Under construction

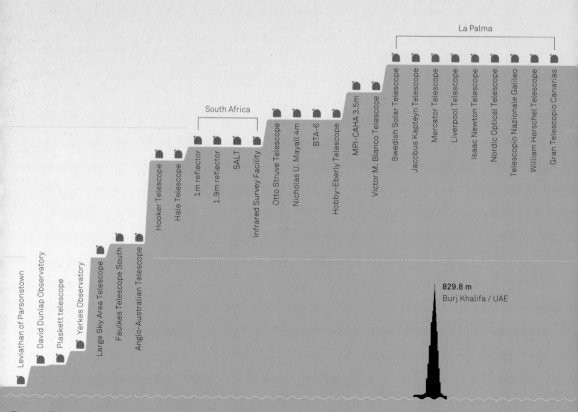

La Palma

Leviathan of Parsonstown
David Dunlap Observatory
Plaskett telescope
Yerkes Observatory
Large Sky Area Telescope
Faulkes Telescope South
Anglo-Australian Telescope
Hooker Telescope
Hale Telescope
1m reflector
1.9m reflector
SALT
Infrared Survey Facility
Otto Struve Telescope
Nicholas U. Mayall 4m
BTA-6
Hobby-Eberly Telescope
MPI-CAHA 3.5m
Victor M. Blanco Telescope
Swedish Solar Telescope
Jacobus Kapteyn Telescope
Mercator Telescope
Liverpool Telescope
Isaac Newton Telescope
Nordic Optical Telescope
Telescopio Nazionale Galileo
William Herschel Telescope
Gran Telescopio Canarias

South Africa

829.8 m
Burj Khalifa / UAE

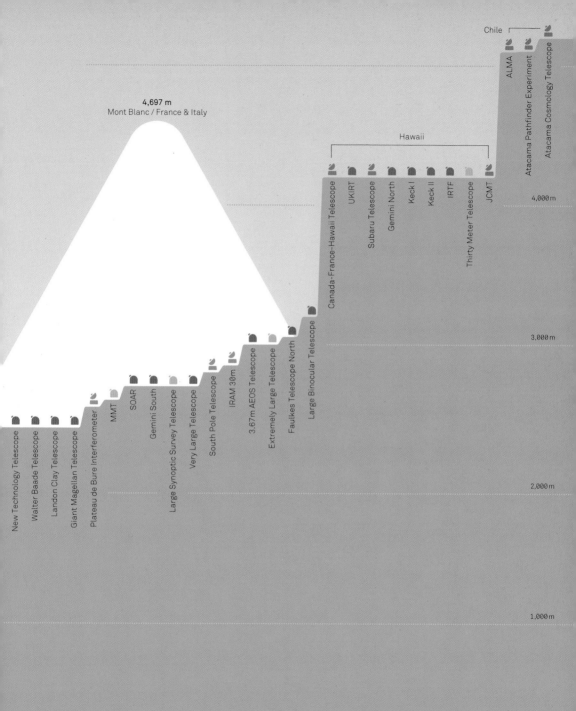

Chile

4,697 m
Mont Blanc / France & Italy

Hawaii

New Technology Telescope
Walter Baade Telescope
Landon Clay Telescope
Giant Magellan Telescope
Plateau de Bure Interferometer
MMT
SOAR
Gemini South
Large Synoptic Survey Telescope
Very Large Telescope
South Pole Telescope
IRAM 30m
3.67m AEOS Telescope
Extremely Large Telescope
Faulkes Telescope North
Large Binocular Telescope
Canada-France-Hawaii Telescope
UKIRT
Subaru Telescope
Gemini North
Keck I
Keck II
IRTF
Thirty Meter Telescope
JCMT
ALMA
Atacama Pathfinder Experiment
Atacama Cosmology Telescope

4,000 m

3,000 m

2,000 m

1,000 m

Sea level

The sky's not the limit

Sitting atop mountains is not the only way of getting above the limiting effects of our atmosphere. To get higher, telescopes have been placed in aeroplanes and hung from high altitude balloons. Even these measures do not completely mitigate the atmospheric effects. The only way to do that is to get into space. Although more expensive to build, and with few opportunities to fix them if they break, humanity now has a fleet of telescopes in orbit.

● Radio　● Infrared / sub-mm　● Optical　● X-ray / Gamma-ray

550km
Fermi Gamma-ray Space Telescope / Low-Earth Orbit

569km
Hubble Space Telescope / Low-Earth Orbit

580 km
Swift / Low-Earth Orbit

10 km　　　　　100 km　　　　　1,000 km　　　　　10,000 km

650 km
SWAS / Low-Earth Orbit

768km
Compton Gamma-ray Observatory / Low-Earth Orbit

13 km
SOFIA / Boeing 747

900km
IRAS / Sun-synchronous Earth Orbit

40km
BLAST / High-altitude balloon

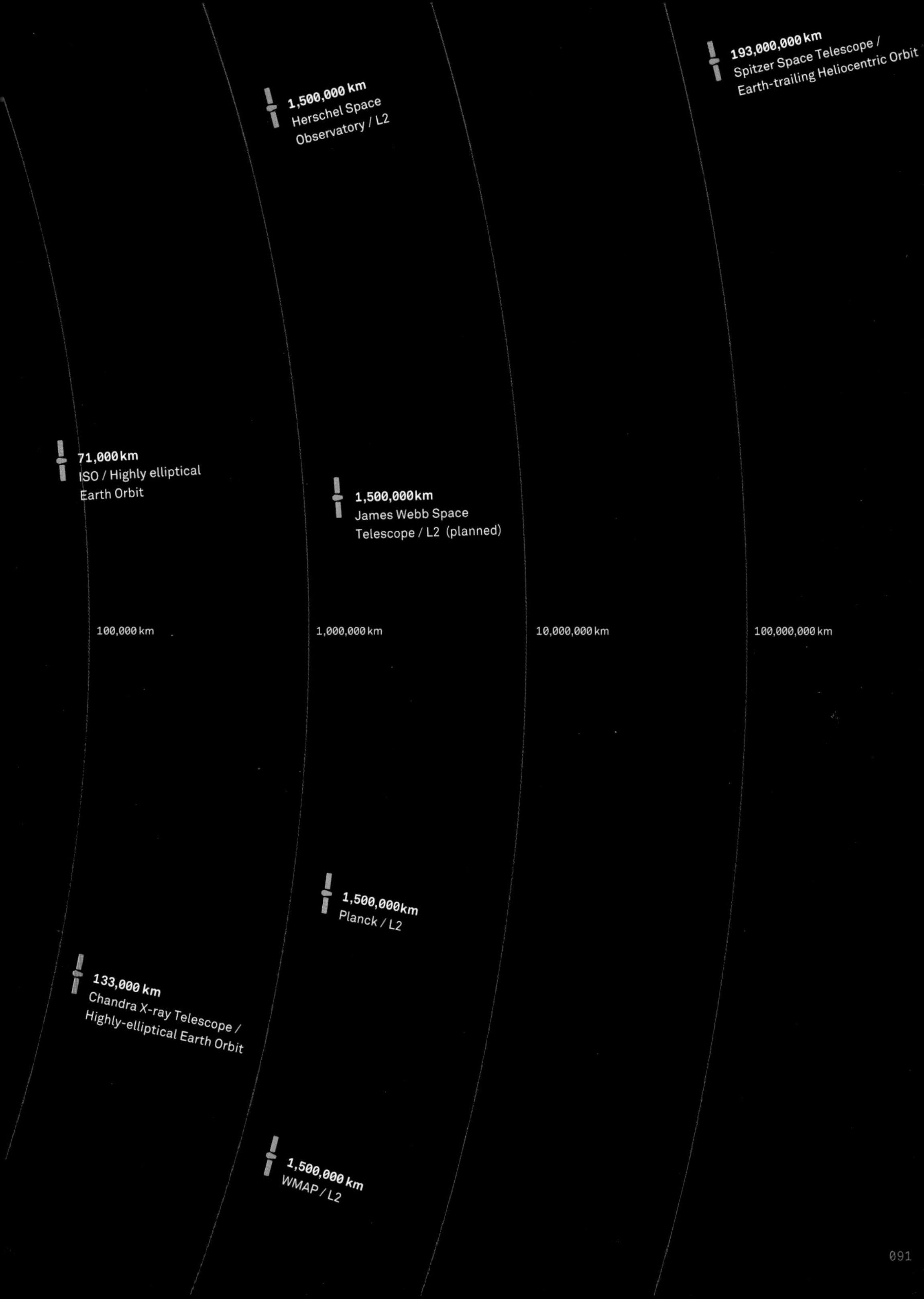

193,000,000 km
Spitzer Space Telescope /
Earth-trailing Heliocentric Orbit

1,500,000 km
Herschel Space
Observatory / L2

71,000 km
ISO / Highly elliptical
Earth Orbit

1,500,000 km
James Webb Space
Telescope / L2 (planned)

100,000 km

1,000,000 km

10,000,000 km

100,000,000 km

1,500,000 km
Planck / L2

133,000 km
Chandra X-ray Telescope /
Highly-elliptical Earth Orbit

1,500,000 km
WMAP / L2

Radio telescopes / bigger and bigger

The light we see with our eyes is only a tiny fraction of the radiation that is available for study. In the early 20th Century astronomers started building telescopes that collect radio waves.

Some are not too unlike optical telescopes, comprising a large dish, acting as the mirror. As with optical telescopes, a larger mirror allows fainter objects to be observed in greater detail, and so there has always been a desire for larger radio telescopes. To achieve even finer detail, many dishes can be joined together to act as one giant telescope.

● **VLBA**
Run by the US National Radio Astronomy Observatory, this spans the United States.

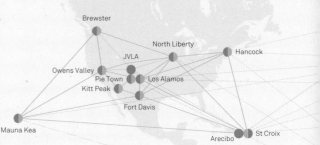

● **Global VLBI**
To get the best resolution, VLBA, EVN, and space-based telescopes occasionally team up to act as one telescope three times the size of the Earth.

Sizes of telescopes

A larger telescope means more detail can be seen. China are building the world's most sensitive radio telescope – Five Hundred Meter Aperture Spherical Telescope (FAST).

9m
Reber's Radio Telescope / US

32m
RT-4 Telescope / Poland

64m
Sardinia Radio Telescope / Italy

110m
Robert C. Byrd Telescope / US

25m
Onsala 25-m / Sweden

38.1m
MKII Telescope / UK

76m
Lovell Telescope / UK

305m
Arecibo / Puerto Rico

26m
HartRAO / South Africa

64m
Parkes / Australia

100m
Effelsberg / Germany

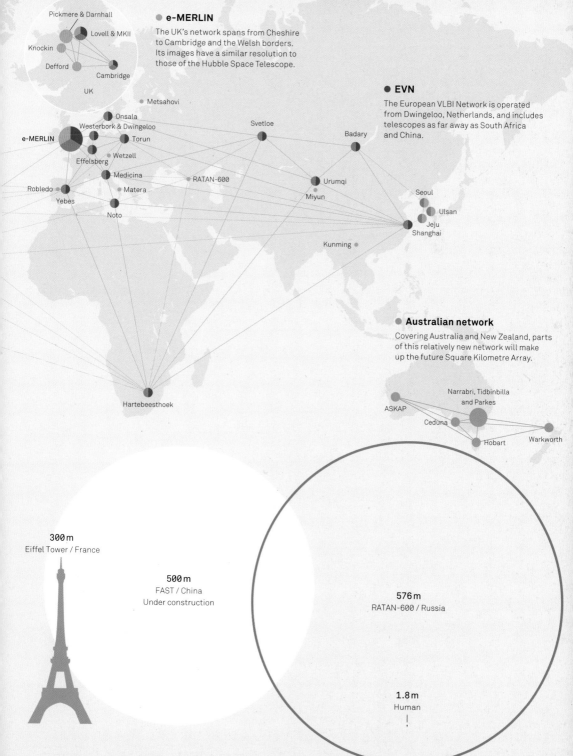

e-MERLIN

The UK's network spans from Cheshire to Cambridge and the Welsh borders. Its images have a similar resolution to those of the Hubble Space Telescope.

Pickmere & Darnhall
Lovell & MKII
Knockin
Defford
Cambridge
UK

EVN

The European VLBI Network is operated from Dwingeloo, Netherlands, and includes telescopes as far away as South Africa and China.

Metsahovi
Onsala
Westerbork & Dwingeloo
e-MERLIN
Torun
Wetzell
Effelsberg
Medicina
RATAN-600
Robledo
Matera
Yebes
Noto
Svetloe
Badary
Urumqi
Miyun
Seoul
Ulsan
Jeju
Shanghai
Kunming

Hartebeesthoek

Australian network

Covering Australia and New Zealand, parts of this relatively new network will make up the future Square Kilometre Array.

ASKAP
Ceduna
Narrabri, Tidbinbilla and Parkes
Hobart
Warkworth

300 m
Eiffel Tower / France

500 m
FAST / China
Under construction

576 m
RATAN-600 / Russia

1.8 m
Human

093

Telescope chronology

To get the best images astronomers seek ever-larger
telescopes. During the 19th Century the largest was
the Leviathan of Parsonstown in Ireland where the
spiral arms of the Andromeda Galaxy were first seen.
The 20th Century saw much larger telescopes built
in the continental US and later Hawaii and Chile.

The first radio telescopes were in the 1930s but it wasn't
until the start of the space race that the first large radio
telescopes were constructed. Since the 1980s, radio
astronomers have linked vast networks of telescopes
together to create a telescope larger than our planet.

Telescope operational period 1840 –

↑ Largest telescope
▬ Radio
▬ Infrared / Sub-mm
▬ Optical
  Planned
↓ Smallest telescope

✎ Located on orbital satellite

Jansky's merry-go-round / US ········

Hale Telescope / US ········

Hooker Telescope / US ▬▬▬▬▬▬▬▬

Otto Struve Telescope / US ········

Leviathan of Parsonstown / Ireland ▬▬▬▬▬▬▬▬▬▬▬▬▬▬▬▬▬

Yerkes Observatory / US ▬▬▬▬▬▬▬▬▬▬▬

| 1850 | 1855 | 1860 | 1865 | 1870 | 1875 | 1880 | 1885 | 1890 | 1895 | 1900 | 1905 | 1910 | 1915 | 1920 | 1925 | 1930 |

space-VLBI / Low-Earth Orbit
VLBA / North America
e-MERLIN / UK
Jansky Very Large Array / US
Giant Metrewave Radio Telescope / India
ALMA / Chile
RATAN-600 / Russia
Arecibo / Puerto Rico
Robert C. Byrd Telescope / US
Effelsberg / Germany
300 Foot Telescope / US
Lovell Telescope / UK
Parkes / Australia

Extremely Large Telescope / Chile
Thirty Meter Telescope / US

JCMT / US
Gran Telescopio Canarias / La Palma
SALT / South Africa
Keck I / US

Reber's radio telescope / US

Large Synoptic Survey Telescope / Chile
Subaru Telescope / US
Very Large Telescope / Chile
Gemini North / US
James Webb Space Telescope / L2

William Herschel Telescope / La Palma
Anglo-Australian Telescope / Australia
UKIRT / US
Canada-France-Hawaii Telescope / US
New Technology Telescope / Chile
ESO 3.6 m Telescope / Chile
Herschel Space Observatory / L2
Isaac Newton Telescope / La Palma

Hubble Space Telescope / Low-Earth Orbit

Faulkes Telescope North / US
Faulkes Telescope South / Australia
Planck / L2

WMAP / L2
Spitzer Space Telescope / Earth-trailing Heliocentric Orbit
IRAS / Sun-synchronous Earth Orbit ●
ISO / Highly-elliptical Earth Orbit
COBE / Sun-synchronous Earth Orbit

1940 1945 1950 1955 1960 1965 1970 1975 1980 1985 1990 1995 2000 2005 2010 2015 2020

Megapixels

Over recent decades digital imaging technology has advanced at an incredible rate, and astronomy has been at the forefront of the development. The detectors are similar in design to those in standard smart phones and digital cameras, but much more sensitive. The standard unit of measurement for a camera is a megapixel (one million pixels).

While early astronomical cameras were very modest, recent developments have allowed cameras to be built with billions of pixels. Most of these are on ground-based telescopes, with the exception of the 938-million-pixel camera on board the Gaia spacecraft. In general, interplanetary spacecraft have much smaller cameras because they were designed and built long before they arrive at their destinations but also because they have limited bandwidth to send images back to Earth.

20 megapixels
35mm film (equivalent)
Non-astronomical

13 megapixels
Canon EOS 5D DSLR
Non-astronomical

8 megapixels
iPhone 6
Non-astronomical

1 megapixel
Early digital Camera
Non-astronomical

938 megapixels Gaia spacecraft / Astronomical
126 megapixels SDSS / Astronomical
95 megapixels Kepler space observatory / Spacecraft
80 megapixels Suprime-cam (Subaru) / Astronomical
36 megapixels LMI (Discovery Channel Telescope) / Astronomical
17 megapixels WFC3 (Hubble) / Astronomical
8 megapixels WFC (INT) / Astronomical
4 megapixels OSIRIS (Rosetta) / Spacecraft
1.9 megapixels MastCam & MAHLI (Curiosity) / Spacecraft
1 megapixel LORRI (New Horizons), MDIS (Messenger) & ISS (Cassini) / Spacecraft

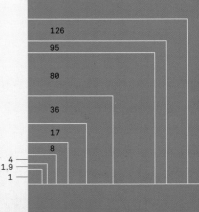

3,200 megapixels
Gigacam (Large Synoptic Survey Telescope) / Astronomical

1,400 megapixels
Panstarrs / Astronomical

938

870 megapixels
Hyper Suprime-Cam / Astronomical

570 megapixels
Dark Energy Camera (M. Blanco telescope) / Astronomical

340 megapixels
Megacam (CFHT) / Astronomical

126

95

80

36

17

8

4
1.9
1

Resolution

The unaided human eye can see objects down to about the size of the tip of a pin held at arm's length. That is equivalent to just being able to make out something 111 kilometres across on the surface of the Moon.

Larger telescopes can see finer detail and this has led to huge advances in astronomy. For optical and infrared telescopes the Earth's atmosphere is often the limiting factor because the turbulent air constantly distorts the light we see on the ground. For radio telescopes there is no such obstacle.

We can use an eye-chart to show the relative detail seen by different instruments, along with the equivalent size of objects they could discern at the distance of the Moon.

Row 4 / 144 arcseconds
Fermi – gamma-ray telescope and size of the **Hubble Ultra Deep Field** image.
Smallest feature on the Moon 265 km

Row 8 / 60 arcseconds
Human eye (20:20 vision).
Smallest feature on the Moon 111 km

Row 1 / 300 arcseconds
Planck – microwave satellite.
Smallest feature on the Moon 553 km

Row 7 / 66 arcseconds
Angular size of Venus (closest).
Smallest feature on the Moon 122 km

Row 9 / 43 arcseconds
Diameter of Saturn's rings at opposition.
Smallest feature on the Moon 79 km

L E F F O D P C F T O A

P C Z O L C D P T O N

L E P Z C L O D T F C F T P

P Z L F T D F N O L T F E E N Z P

F P C F L F Z O O P L F B E L F P D F

Row 11 / 25 arcseconds
Angular size of Mars (closest).
Smallest feature on the Moon 46 km

Row 13 / 18 arcseconds
Herschel – infrared satellite.
Smallest feature on the Moon 33 km

Row 15 / 9.5 arcseconds
Angular size of Venus (furthest).
Smallest feature on the Moon 18 km

Row 20 / 3.5 arcseconds
Angular size of Mars (furthest).
Smallest feature on the Moon 6.4 km

Row 22 / 2 arcseconds
Angular size of Neptune.
Smallest feature on the Moon 3.7 km

Row 24 / 1.2 arcseconds
90 mm (3.5 inch) backyard telescope.
Smallest feature on the Moon 2.2 km

Row 25 / 1 arcsecond
Blurring effect of the atmosphere
at dark sites. Smallest feature on
the Moon 1.8 km

Row 27 / 0.7 arcseconds
Cassini division in Saturn's rings.
Smallest feature on the Moon 1.3 km

Row 28 / 0.5 arcseconds
Keck telescope.
Smallest feature on the Moon 920 m

Row 29 / 0.4 arcseconds
Blurring effect of the atmosphere
at dark, high altitude sites.
Smallest feature on the Moon 735 m

Row 33 / 0.16 arcseconds
JWST – infrared satellite.
Smallest feature on the Moon 295 m

Row 35 / 0.1 arcseconds
Angular size of Pluto.
Smallest feature on the Moon 184 m

Row 38 / 0.05 arcseconds
Hubble Space Telescope and the angular
size of Betelgeuse.
Smallest feature on the Moon 92 m

Row 39 / 0.04 arcseconds
e-MERLIN – radio interferometer.
Smallest feature on the Moon 74 m

Row 45 / 0.01 arcseconds
Keck telescope theoretical resolution.
Smallest feature on the Moon 18 m

Row 55 / 0.001 arcseconds
VLTI – optical interferometer.
Smallest feature on the Moon 1.8 m
(e.g. a human)

Row 63 / 0.00015 arcseconds
EVN – radio interferometer.
Smallest feature on the Moon 0.3 m

Four / The Sun

The Sun

The Sun is the nearest star to Earth. It is a huge ball
of plasma with thermonuclear reactions in its core.
The light and heat it generates sustains almost all life
on our planet. It is about 4.567 billion years old and
is half way through its life.

Total light output
383,000,000,000,000,000,000,000,000 watts
(383 million billion billion watts)

Mass
1,989,000,000,000,000,000,000,000,000,000 kg
(1,989 billion billion billion kg) = 330,000 times Earth

Mass loss
620,000,000,000 kg/s (620 billion kg)

Rotation at poles 36 days

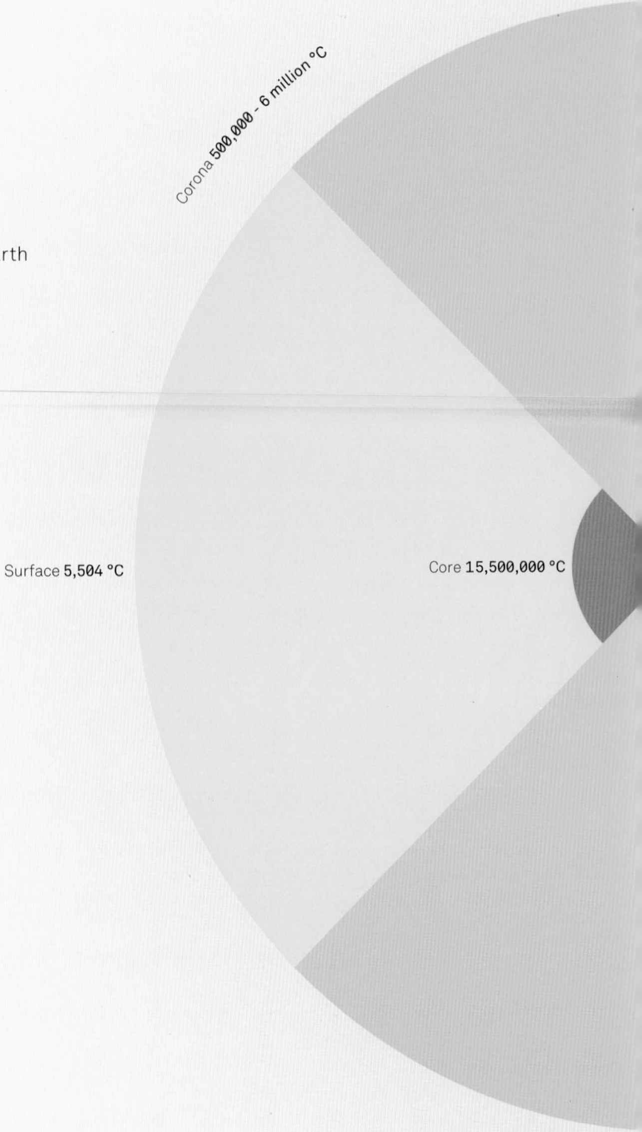

Corona 500,000 - 6 million °C

Surface 5,504 °C

Core 15,500,000 °C

Rotation equator **26.8 days**

Time for light to reach the surface
150,000 - 1 million years

Time for light to reach the Earth
from surface **8.3 minutes**

Earth

Spectrum of the Sun

The Sun's light can be split up into a spectrum of colours, most often seen as a rainbow. In the 19th Century, astronomers studying this rainbow realised that it contained dark bands. In 1860 Gustav Kirchoff and Robert Bunsen discovered that each chemical element produces its own particular set of bands at specific colours – a spectral fingerprint.

While analysing the spectrum of the Sun in 1868, astronomers Jules Janssen and Norman Lockyear identified a previously unknown element. It took until 1895 before Per Teodor Cleve and Nils Abraham Langlet finally discovered the new element on Earth. It was named helium after Helios, the Greek god of the Sun. Helium is now known to be the second most abundant element in the Sun, and the Universe as a whole.

As well as hydrogen and helium, the Sun's spectrum shows the presence of a wide range of elements – as well as oxygen in the Earth's atmosphere. The observed elements are in the Sun's upper atmosphere, and were mostly formed by a previous generation of stars.

Ba / Barium
Ca / Calcium
Cr / Chromium
Fe / Iron
H / Hydrogen
He / Helium
Hg / Mercury
Mg / Magnesium
Na / Sodium
O / Oxygen (in Earth's atmosphere)
Sr / Strontium
Ti / Titanium

O

Ca

Ca

Fe

Hg

Fe

Fe —

H

H

Ca

H

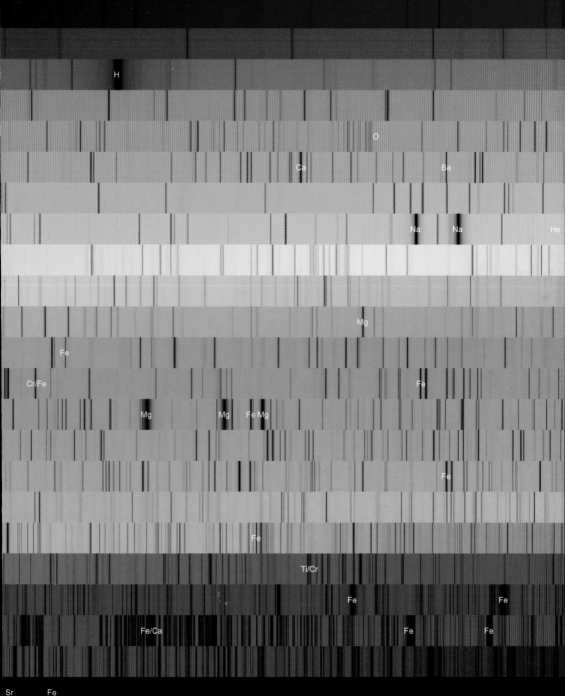

Sunspots

The surface of the Sun is a broiling mass of plasma
and magnetic fields. Where the magnetic field punches
out of the surface, the temperature is kept slightly lower
and less light is produced than elsewhere on the Sun.
These darker patches are called sunspots and they
come and go over time. The overall number of sunspots
changes over an 11 year cycle.

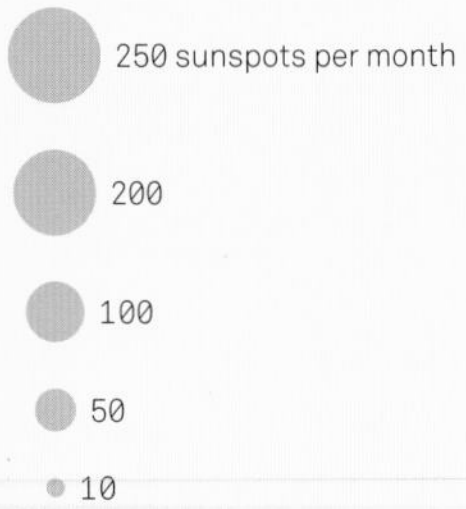

250 sunspots per month

200

100

50

10

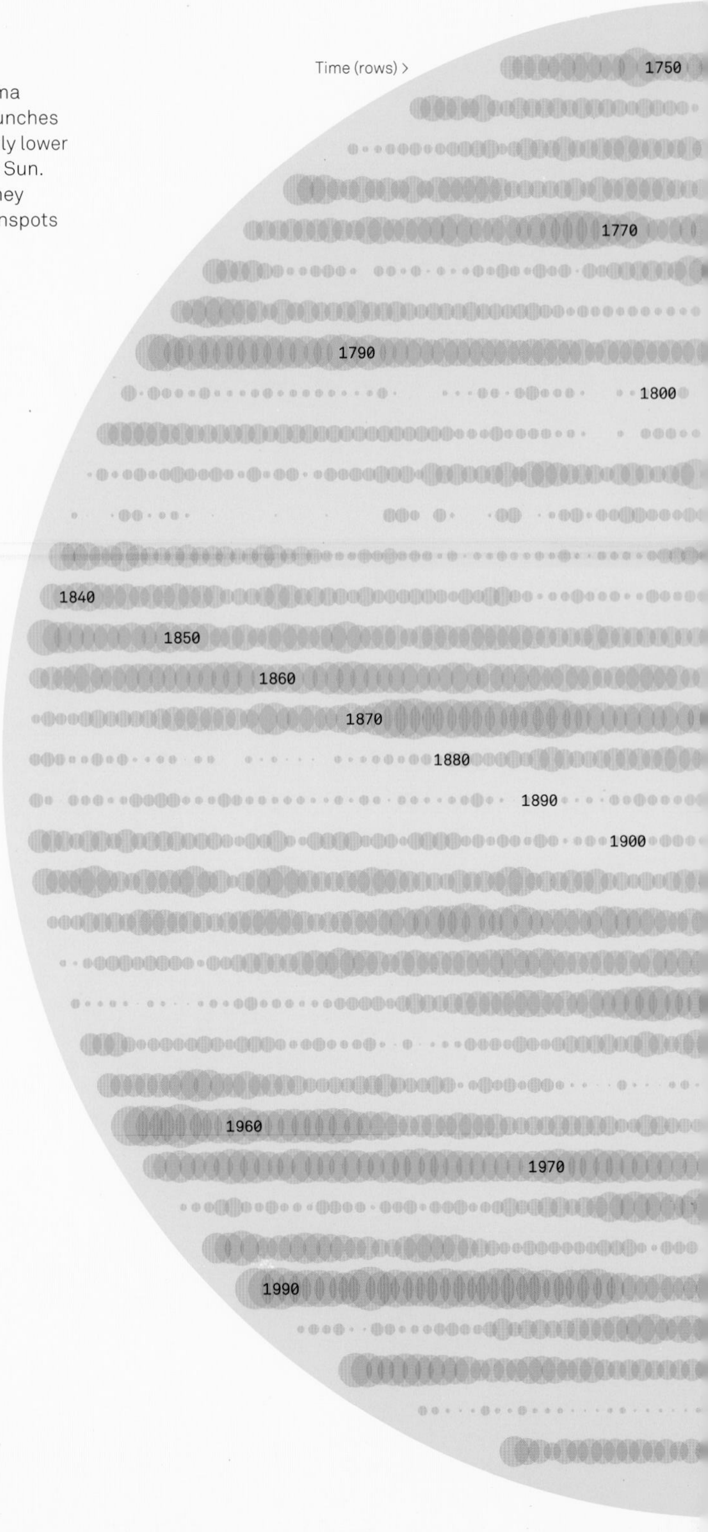

1750

1770

1790

1800

1840

1850

1860

1870

1880

1890

1900

1960

1970

1990

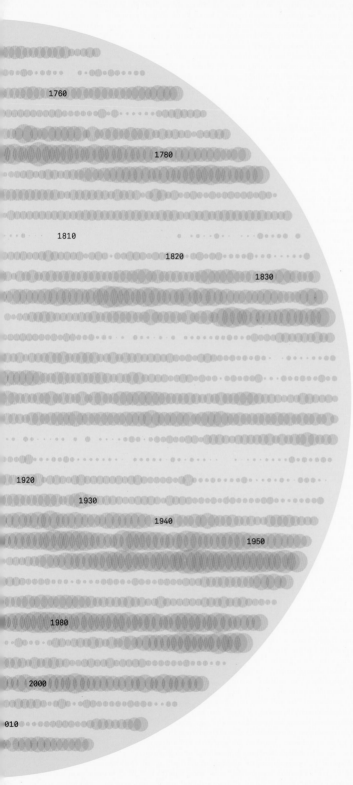

Butterfly diagram

Sunspots don't appear just anywhere on the surface of the Sun. Over the period of approximately 11 years they form gradually closer and closer to the Sun's equator. There is a clear link with the cycle of solar activity, which also takes around 11 years.

In the early 20th Century it was discovered that sunspots are magnetic phenomena, and they highlight where the

Sun's magnetic field is breaking through its surface. Sunspots in alternating cycles have magnetic fields with opposite polarity. This showed that the Sun's magnetic field actually fluctuates over a period of 22 years.

This link between sunspots and the magnetic field, combined with historical data, means that the behaviour of the magnetic cycle can be inferred for hundreds of years.

1960   1970   1980

North pole

60°

0°

-30°

South pole

Sunspots per month

- 1-5
- 5-10
- 10-20
- > 20

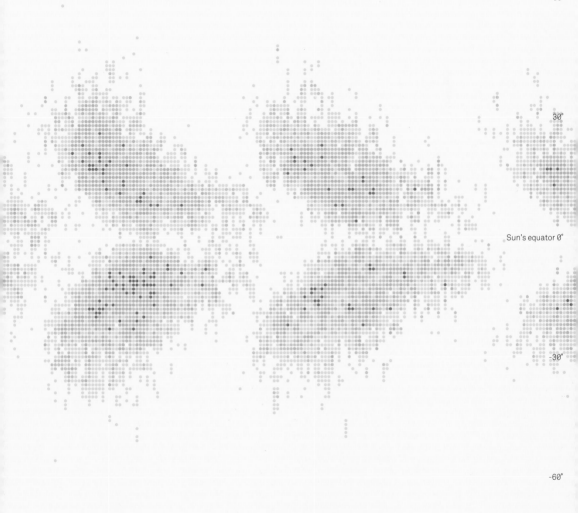

60°

30°

Sun's equator 0°

-30°

-60°

1990   2000   2010

Solar flares

Since the late 1970s satellites have been recording explosive solar flares from the surface of the Sun. These flares are measured and put into several classes – A, B, C, M, or X – where each class is 10 times stronger than the one before. Within each class, the numbers 1 – 9 are used e.g. an M5 flare is five times more powerful than an M1.

An X1 flare is equivalent to around 200 million megatons of TNT or a million times as much energy as a volcanic explosion. There is currently no class above X so the largest flares keep the 'X' designation but increase the number.

C – class
Don't noticeably affect us.

M – class
Can cause radio blackouts near the poles and radiation storms that can affect astronauts.

X – class
Can knock out satellites, increase the radiation doses of airline passengers, and create blackouts in electricity grids on the ground.

Solar cycles

24 / Started January 2008
23 / May 1996 – January 2008
22 / September 1986 – May 1996

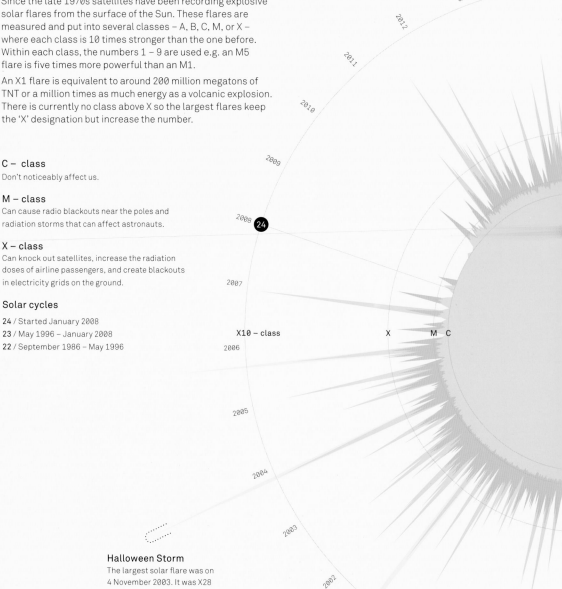

X10 – class

X   M   C

Halloween Storm
The largest solar flare was on 4 November 2003. It was X28

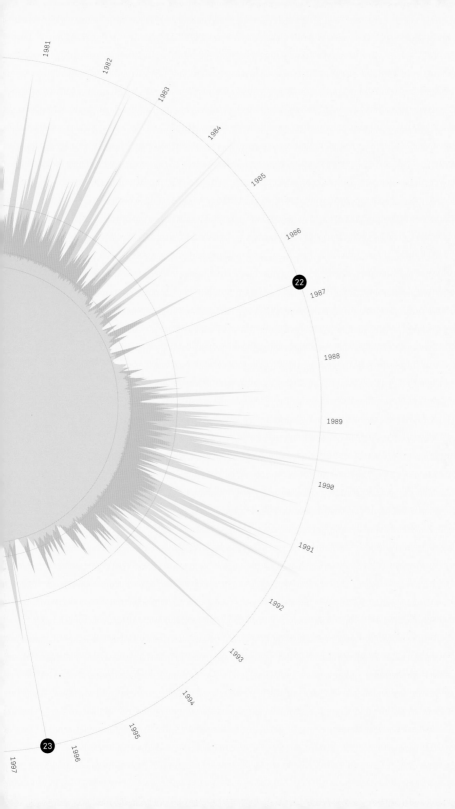

1981
1982
1983
1984
1985
1986
22
1987
1988
1989
1990
1991
1992
1993
1994
1995
23
1996
1997

Dance of the Sun

We often think of the Sun as the centre of the Solar System
but it isn't, not quite. The gravity of its many planetary
partners leads it in complicated loops around the dance
floor of the Solar System.

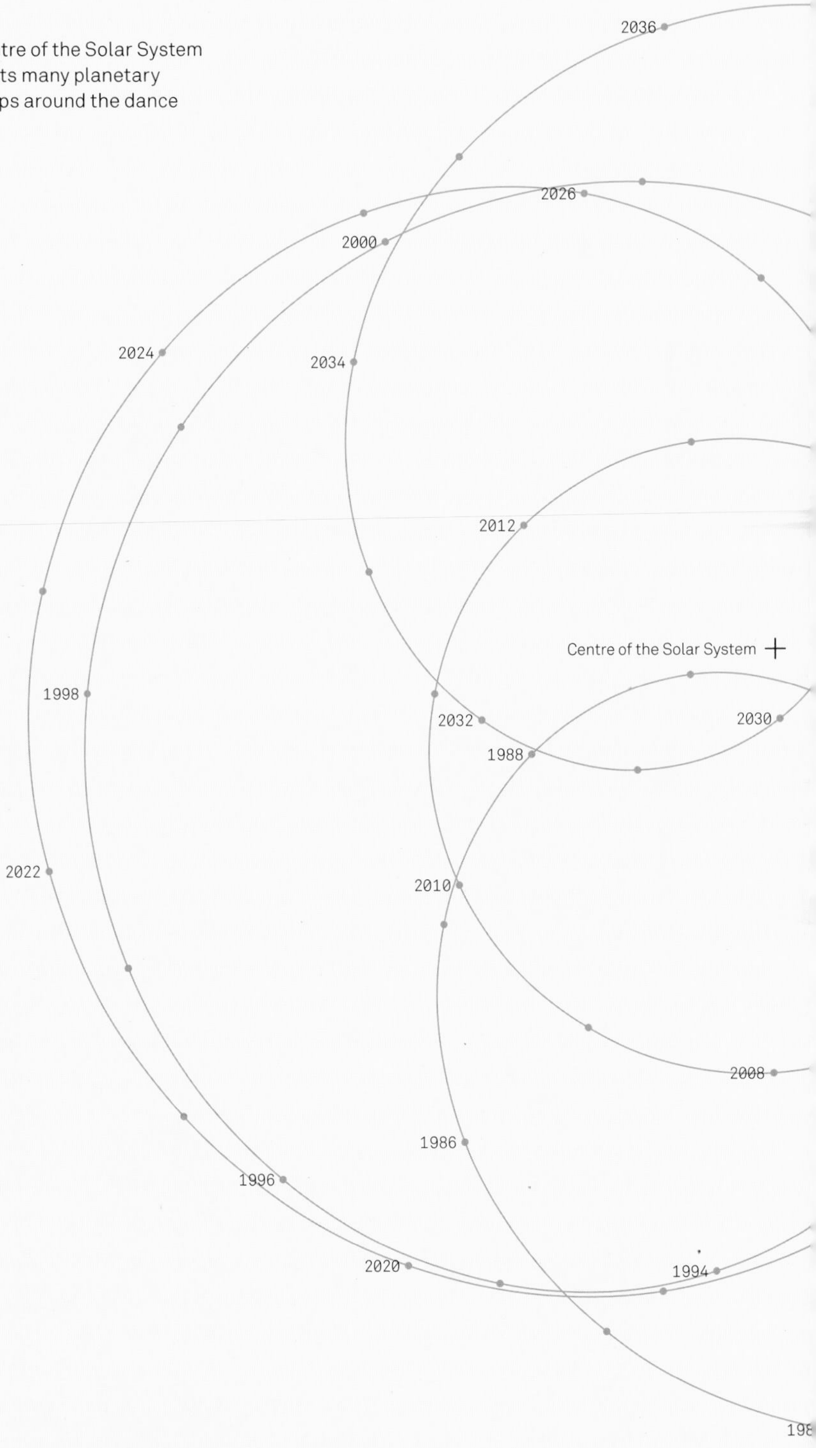

2036

2026

2000

2024

2034

2012

Centre of the Solar System +

1998

2032   2030

1988

2022

2010

2008

1986

1996

2020   1994

198

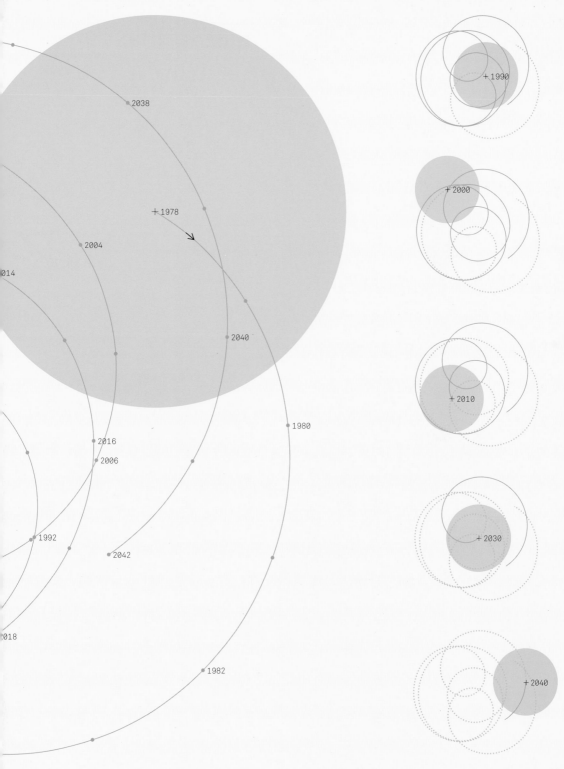

Five / Stars

Northern constellations

To ancient mariners and nomadic peoples, the night sky was a vital tool for navigation. It provided a point of reference for the time of year and latitude on the Earth when the Sun had set. Over the years people linked nearby stars to form patterns. These patterns are called constellations and neighbouring constellations are usually linked together in mythological stories. Today these stories are entertaining but they also help us remember the patterns that could be vital for planting crops, finding our way home, or even across an ocean.

Many constellations have origins in ancient mythology. For instance, the hero Perseus is riding Pegasus (The Winged Horse) – to save Princess Andromeda, daughter of Queen Cassiopeia. Others are named after animals, such as Leo the Lion and Camelopardalis the giraffe (or the camel-leopard!).

One of the most famous northern constellations is Ursa Major (The Great Bear). This contains the famous asterism The Plough, or Big Dipper, as the tail and back of the bear.

Cassiopeia / 598 sq deg

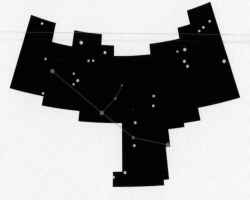

Andromeda / 722 sq deg

Boötes / 907 sq deg

Triangulum / 132 sq deg

Camelopardalis / 757 sq deg

Ursa Minor / 256 sq deg

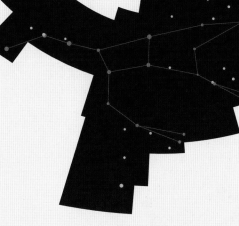

Ursa Major / 1,280 sq deg

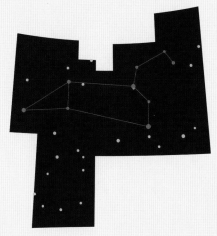

Leo / 947 sq deg

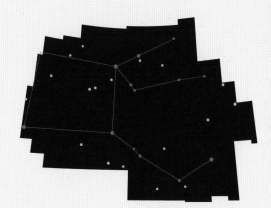

Pegasus / 1,121 sq deg

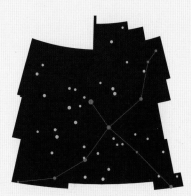

Cygnus / 804 sq deg

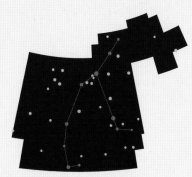

Perseus / 615 sq deg

Southern constellations

The northern skies contain many well known constellations but the southern skies contain the largest and smallest constellations by area; Hydra (Female Water Snake) and Crux (The Southern Cross).

The names of the southern constellations often have more contemporary origins than their northern counterparts e.g. Norma (The Set Square, or Ruler), Dorado (Dolphinfish or Swordfish), and Pyxis (The Mariner's Compass).

Crux / 68 sq deg

Sagittarius / 867 sq deg

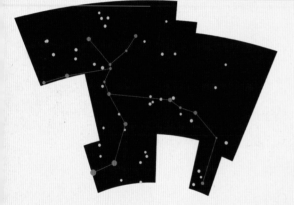

Capricornus / 414 sq deg

Centaurus / 1,060 sq deg

Vela / 500 sq deg

Norma / 165 sq deg

Pyxis / 221 sq deg

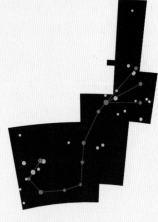

Scorpius / 497 sq deg

Hydra / 1,303 sq deg

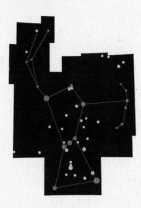

Orion / 594 sq deg

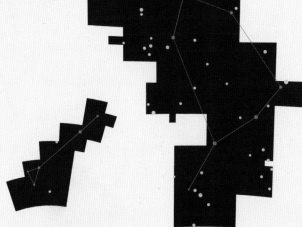

Dorado / 179 sq deg

Ophiuchus / 948 sq deg

Grus / 366 sq deg

3D Orion

Looking at the night sky it is natural to think of it as a tremendous crystal sphere with all the stars placed as points of lights on it. This was how many of the ancients imagined things to be. In reality, all the stars are at different, albeit enormous, distances. Space is 3D and we only see a 2D representation of it.

The familiar constellation of Orion, the Hunter, looks very different when viewed from a different direction. We see that the close grouping is lost when seen from other angles. Even the three stars of the belt turn out to contain an interloper in the form of the central star Alnilam which is a much brighter star but twice as far from Earth.

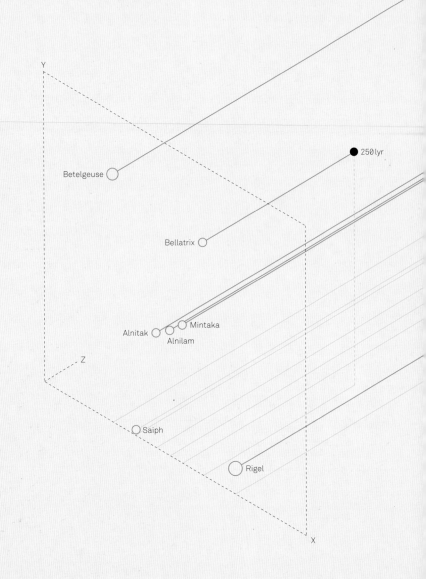

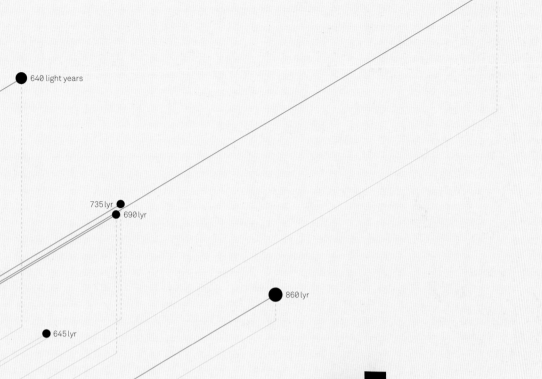

640 light years

1,340 lyr

735 lyr
690 lyr

860 lyr

645 lyr

The position of Orion on the sky makes it visible to all cultures around the world. To the Greeks, Orion was a mighty Hunter. He was accompanied by his dogs (Canis Major and Canis Minor) as he fought Taurus (The Bull). In African bush stories, the three stars of Orion's belt are zebra being hunted by the bright star Aldebaran. In Australian aboriginal astronomy the belt is known as Djulpan (Three Brothers in a Canoe) and warns of the Monsoon season.

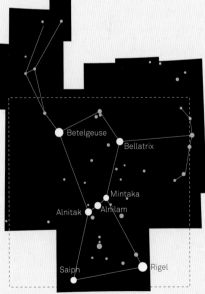

Betelgeuse

Bellatrix

Mintaka

Alnitak
Alnilam

Saiph

Rigel

Orion / 594 sq deg

Nearest stars

Which are the closest stars to Earth? Aside from the Sun, our nearest neighbours are the Alpha Centauri triple star system about 4.3 light years away (40.7 trillion kilometres). The system contains two closely orbiting stars *Alpha Centauri A* and *B* orbited by Proxima Centauri. Stars are not static though and their positions change over time.

Barnard's star is moving towards us relatively quickly and will pass within four light years around 10,000 years from now. In approximately 35,000 years time the star *Ross 248* will be the closest star at around three light years. In roughly 40,000 years time, the *Voyager 1* spacecraft will be as close as 1.6 light years to the star *Gliese 445*.

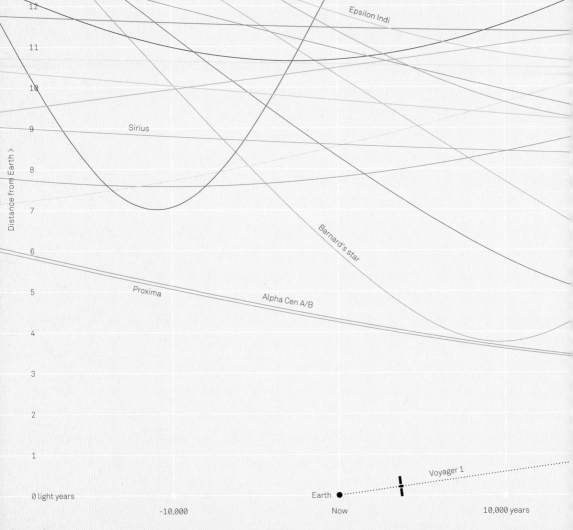

Kapteyn's Star

Epsilon Indi

Sirius

Barnard's star

Proxima

Alpha Cen A/B

Voyager 1

Earth

Distance from Earth >

14
13
12
11
10
9
8
7
6
5
4
3
2
1
0 light years

-10,000   Now   10,000 years

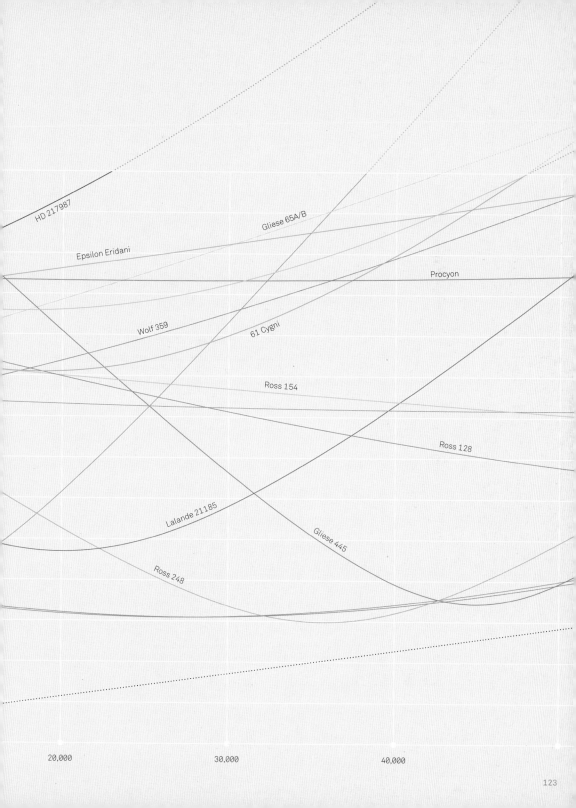

HD 217987

Gliese 65A/B

Epsilon Eridani

Procyon

Wolf 359

61 Cygni

Ross 154

Ross 128

Lalande 21185

Gliese 445

Ross 248

20,000

30,000

40,000

Proper motions

Throughout history astronomers have referred to the 'fixed stars' but, in reality, the stars are not static. Most stars are moving at different speeds and in different directions, though some move together through space. Given the huge distances involved, even at tens of kilometres per second, their motions are imperceptable to our eyes. With careful observations this 'proper motion' across the sky can be measured and, by making calculations, we can see what the sky will look like far off into the future. The patterns in the sky will look different to our descendants.

In 100,000 years from now the constellation of Leo will have lost its couchant pose, while the Gemini twins will have been decapitated! Cassiopeia will no longer be the familiar 'W' shape and Canis Major will have lost the 'Dog Star' – Sirius – from its collar. Although five of the bright stars in 'The Plough' – part of Ursa Major – are moving together through space, the bear will end up with a kink in its tail. The constellation of Orion will have adjusted his sword and shield but one of his shoulders – Betelgeuse – may also have ended its life in a supernova explosion.

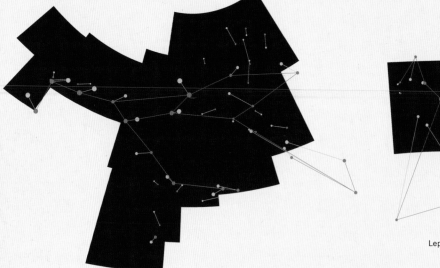

Ursa Major

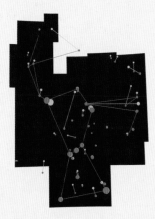

Lepus

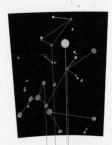

Canis Major

Orion

● Position A, present day

● Position B, 100,000 years in the future

Gemini

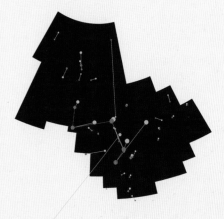

Cassiopeia

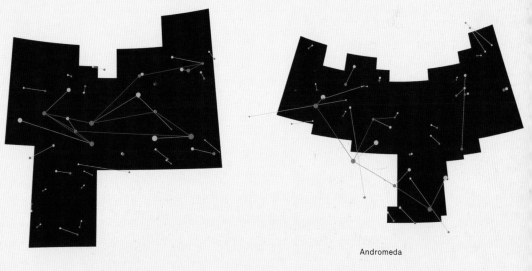

Leo

Andromeda

Brightest stars

When you look up into the night sky you see that some stars are brighter than others. The brightest star in the sky is Sirius, the Dog Star, which is actually a binary star system consisting of Sirius A and Sirius B. The next brightest star is Canopus which is roughly half as bright. Surprisingly, for many, the North Star (Polaris) is far down the list; its importance stems from its position near the pole rather than its brightness. The brightness we see from Earth is actually a combination of a star's true brightness and how close it is. A relatively faint star can appear brighter than others if it is much closer. For example, Canopus is actually 600 times brighter than Sirius, but nearly 40 times further away and so actually appears slightly fainter.

North

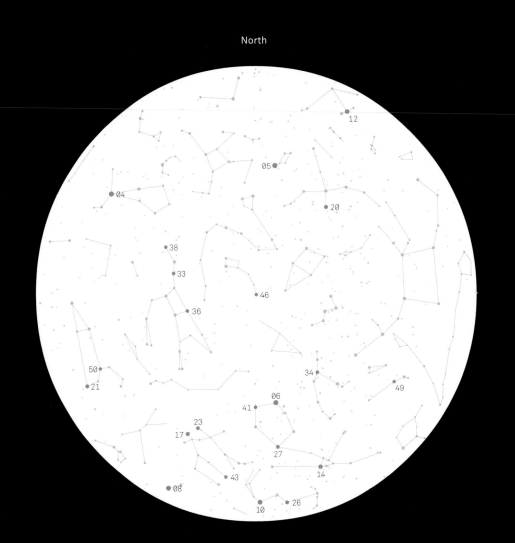

01 / Sirius
02 / Canopus
03 / Alpha Centauri
04 / Arcturus
05 / Vega
06 / Capella
07 / Rigel
08 / Procyon
09 / Achernar
10 / Betelgeuse

11 / Hadar
12 / Altair
13 / Acrux
14 / Aldebaran
15 / Spica
16 / Antares
17 / Pollux
18 / Fomalhaut
19 / Becrux
20 / Deneb

21 / Regulus
22 / Adhara
23 / Castor
24 / Gacrux
25 / Shaula
26 / Bellatrix
27 / Alnath
28 / Miaplacidus
29 / Alnilam
30 / Alnair

31 / Alnitak
32 / Gamma Velorum
33 / Alioth
34 / Mirfak
35 / Epsilon Sagittarii
36 / Dubhe
37 / Wezen
38 / Alkaid
39 / Avior
40 / Theta Scorpii

41 / Menkalinan
42 / Atria
43 / Alhena
44 / Delta Velorum
45 / Peacock
46 / Polaris
47 / Mirzam
48 / Alphard
49 / Hamal
50 / Algieba

South

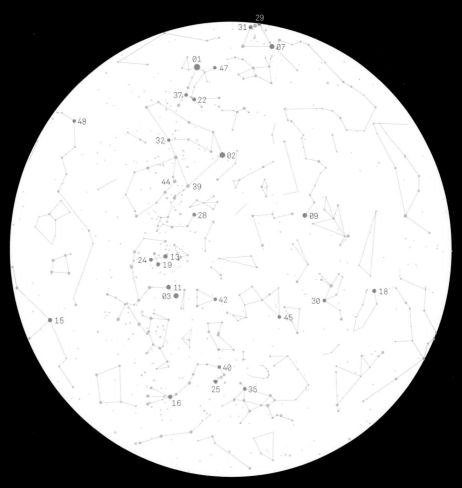

Giant stars

Our local star, the Sun, is roughly 1.4 million kilometres across. That is large – over 100 times the diameter of the Earth – but is tiny compared to many others. One of the contenders for largest known star is *UY Scuti*, which is found in the southern constellation of Scutum. At an estimated 1,700 times the size of the Sun, it would stretch beyond the orbit of Jupiter if it is placed at the centre of the Solar System.

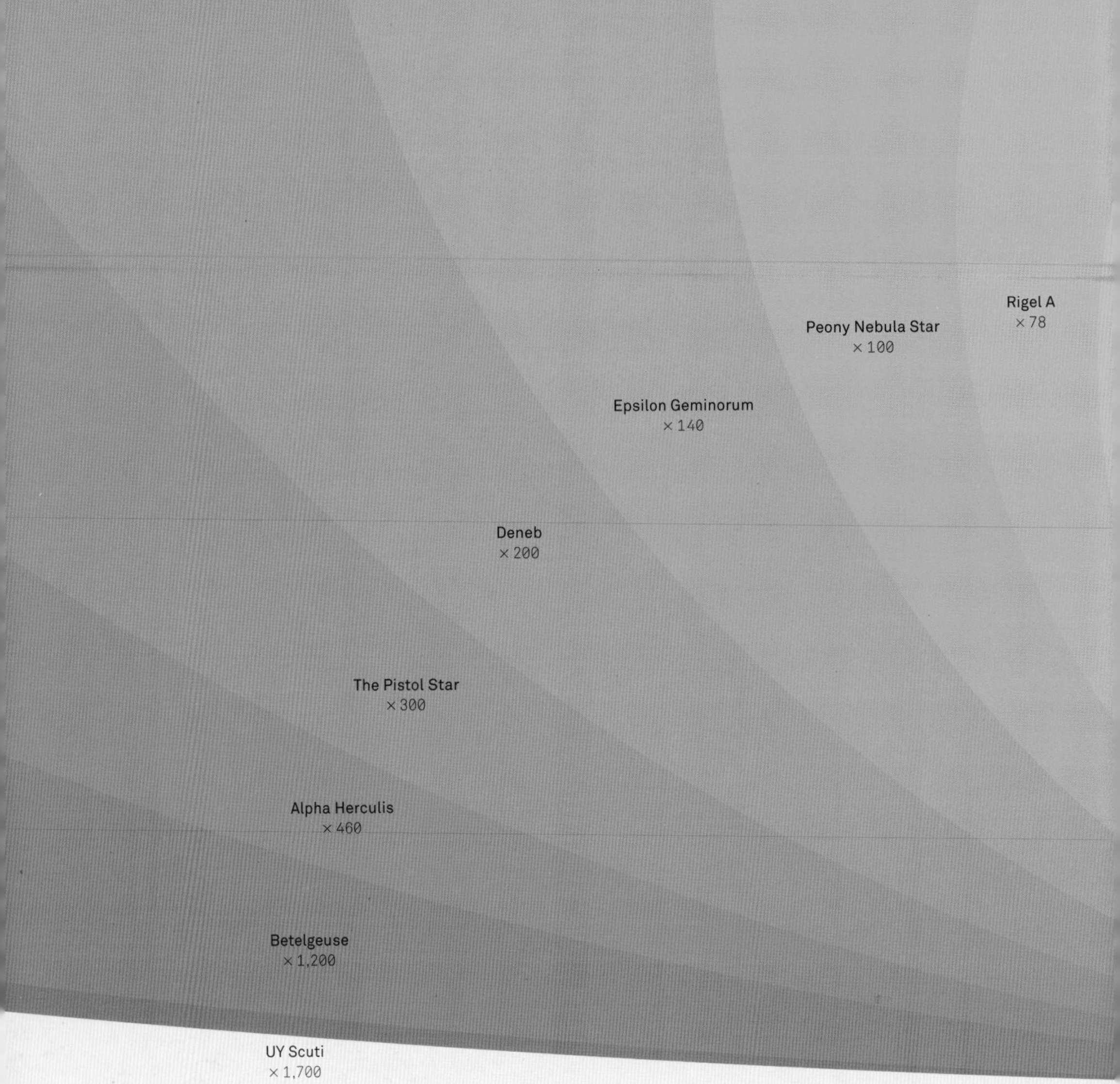

Rigel A
× 78

Peony Nebula Star
× 100

Epsilon Geminorum
× 140

Deneb
× 200

The Pistol Star
× 300

Alpha Herculis
× 460

Betelgeuse
× 1,200

UY Scuti
× 1,700

Canopus
× 65

Aldebaran
× 44

Arcturus
× 25

Delta Boötis
× 10 the Sun

Sun

Dwarf stars

Just how small can a star be? A star is usually defined
as a ball of plasma, held together by its own gravity, and
shining due to thermonuclear fusion occuring at its core.
For the fusion to occur, the core needs to be extremely hot
and dense. We think there needs to be at least 7% of the
mass of the Sun for there to be enough gravity to make
athe conditions right.

The smallest star currently known is only 8.6% the diameter
of the Sun, one eight-thousandth as bright, and has a
temperature of only 2,100 Kelvin. It is known by the rather
cumbersome designation *2MASS J0523-1403*.

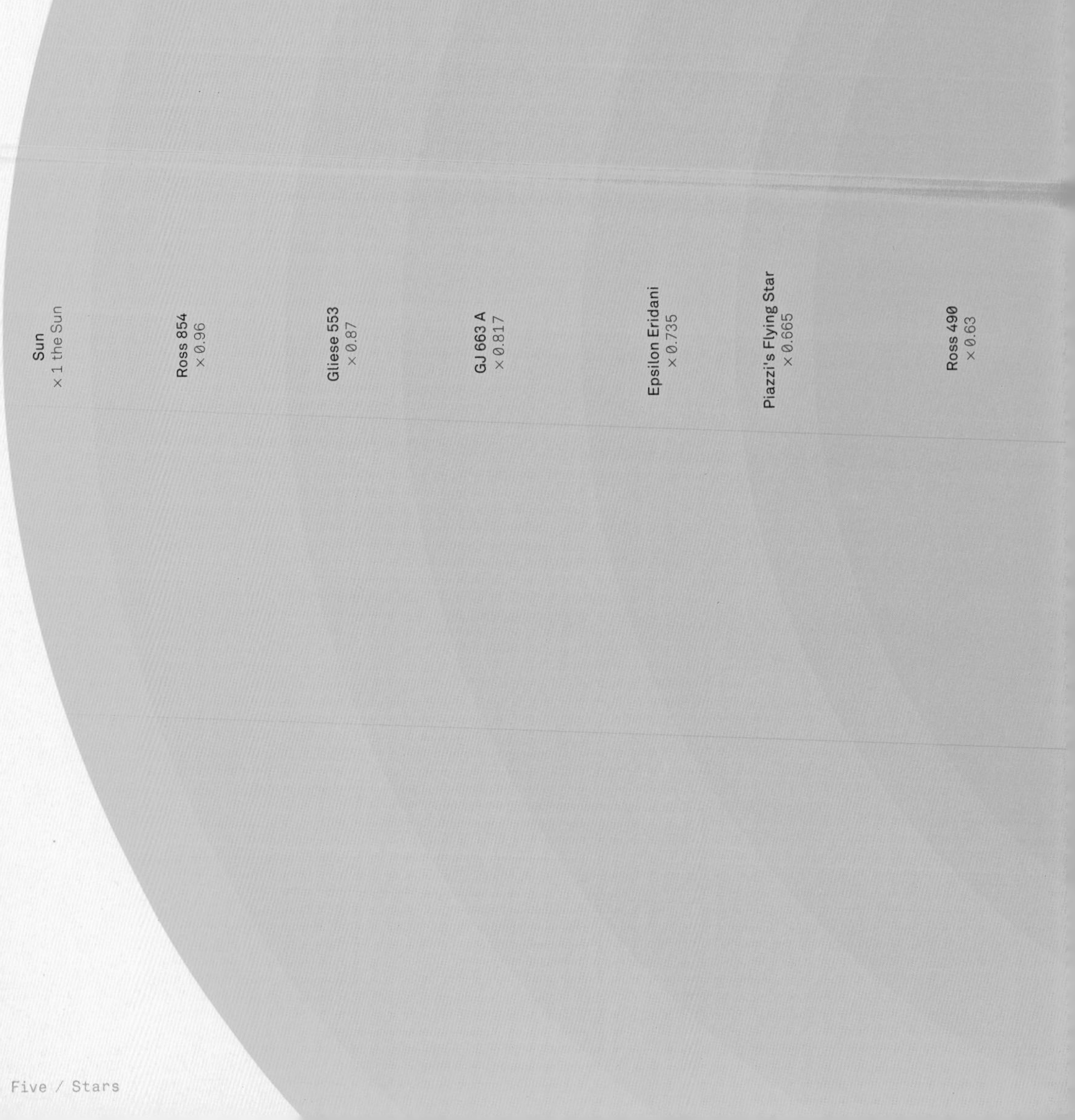

Sun
× 1 the Sun

Ross 854
× 0.96

Gliese 553
× 0.87

GJ 663 A
× 0.817

Epsilon Eridani
× 0.735

Piazzi's Flying Star
× 0.665

Ross 490
× 0.63

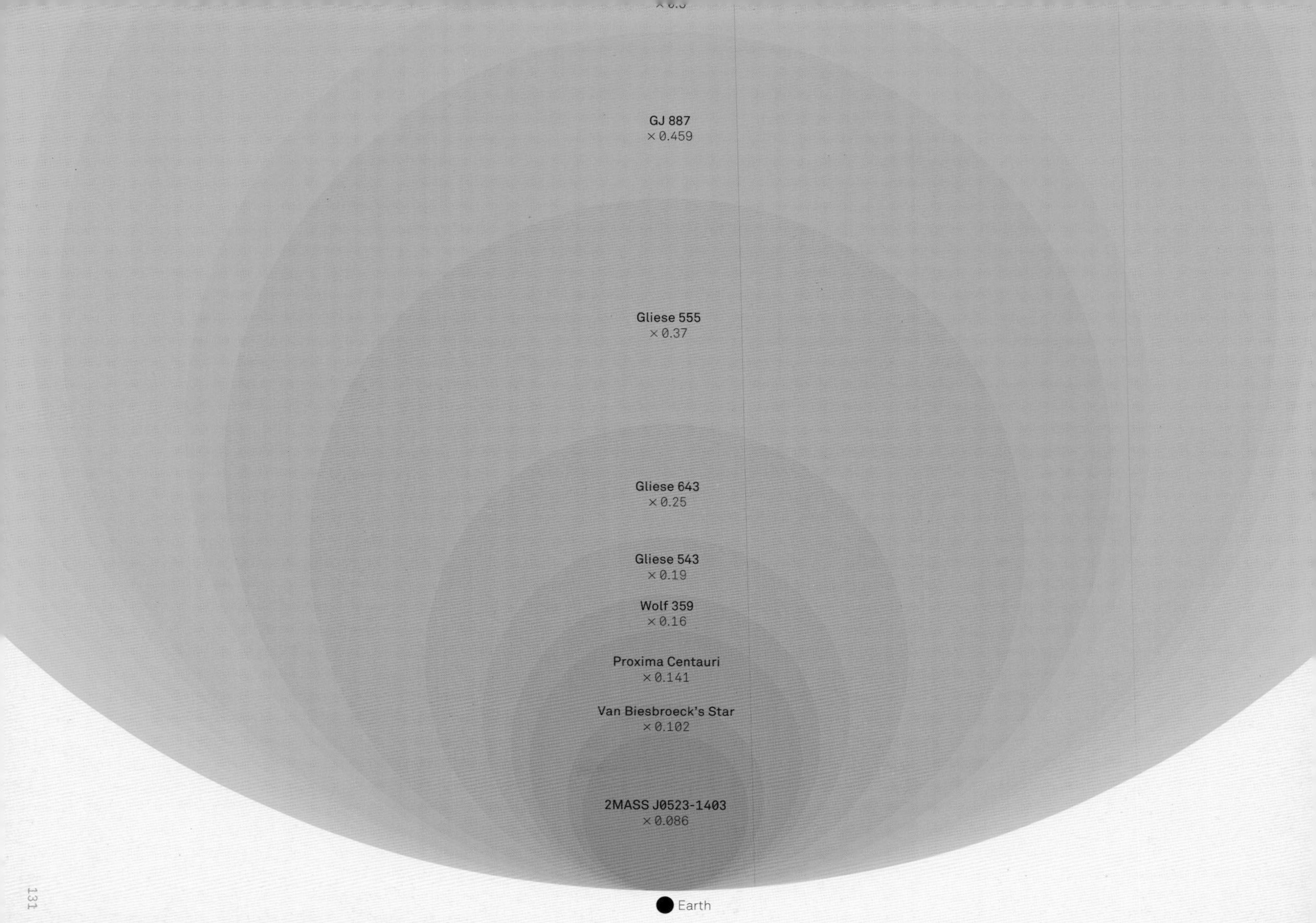

× 0.5

GJ 887
× 0.459

Gliese 555
× 0.37

Gliese 643
× 0.25

Gliese 543
× 0.19

Wolf 359
× 0.16

Proxima Centauri
× 0.141

Van Biesbroeck's Star
× 0.102

2MASS J0523-1403
× 0.086

● Earth

Stellar classifications

In the late 19th and early 20th Century astronomers classified stars into types based on the patterns of dark lines seen in their spectra. The modern classifications were devised by Annie Jump Cannon in 1901, and are given the letters: O, B, A, F, G, K, M. The ordering of the classifications depends on the relative strengths of the various lines, which are caused by the abundance of elements in the star's atmosphere. To remember them, astronomers use the mnemonic 'Oh, Be A Fine Girl/Guy, Kiss Me'. It was only later that it was realised that the order also depends on the surface temperature of the star. The cooler stars have more absortion lines because their cooler atmospheres allow simple molecules to form.

Class / Temperature, Kelvin

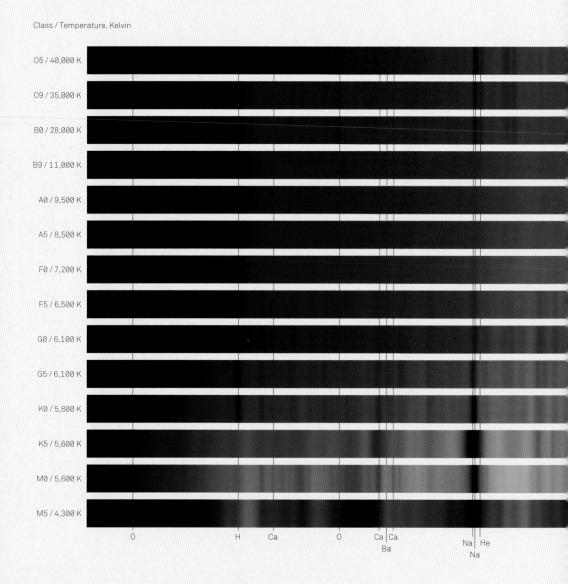

O / Oxygen in Earth's atmosphere

Fe / Iron

H / Hydrogen

Mg / Magnesium

Ca / Calcium

Hg / Mercury

Ba / Barium

Cr / Chromium

Na / Sodium

Ti / Titanium

He / Helium

Sr / Strontium

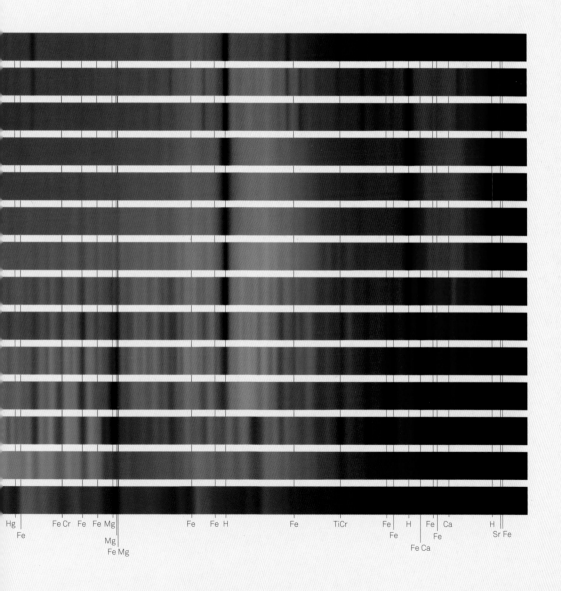

Brightness and colour

In the early 20th Century, Ejnar Hertzsprung and Henry Russell compared the true brightness of stars with their colour. The colour of a star depends on its surface temperature, with hotter stars appearing blue and cooler stars appearing red. By plotting these two properties on a diagram they found that the stars split up into different groups based on how large they were and where they were in their lives. We can make a modern version of a Hertzsprung-Russell diagram using up-to-date measurements for thousands of stars.

Running diagonally across the diagram is the 'Main Sequence' where most stars spend the majority of their lives. Where they are on this sequence depends on their mass at birth. As they age stars expand. This makes them cooler but also brighter, so they move towards the giant and supergiant areas. Just before the very end of their lives, the temperature suddenly increases moving them to the left. After it dies a star will slowly cool and fade as a white dwarf, a neutron star, or maybe even a black hole.

The Sun's journey

A 0 - Sun begins its life.

B 4.5 billion years – The Sun now. Over billions of years the Sun will get slightly brighter and hotter.

C 9.5 billion years – The Sun is expanding into a red giant. It is 2.3 times bigger and 3.2 times brighter.

D 10.3 billion years – The Sun is now 210 times bigger, about 4,200 times brighter, and nearly half its current temperature.

E 10.3 billion years – Having run out of fuel, the Sun has thrown off its outer layers. Just over half of the current mass remains and it has collapsed to about 20% of the original size. Its temperature is around 100,000 Kelvin and it is 3,000 times brighter than it is now.

F 12 billion years – The remnant of the Sun is 0.003% of its current brightness and only about 1.5 times the size of the Earth.

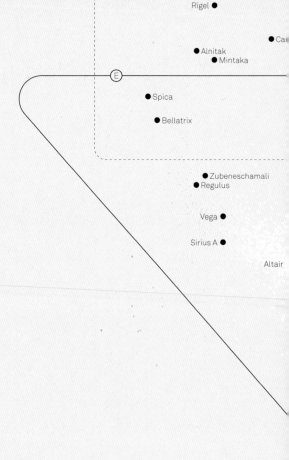

Rigel ●

● Ca

● Alnitak
● Mintaka

E

● Spica

● Bellatrix

● Zubeneschamali
● Regulus

Vega ●

Sirius A ●

Altair

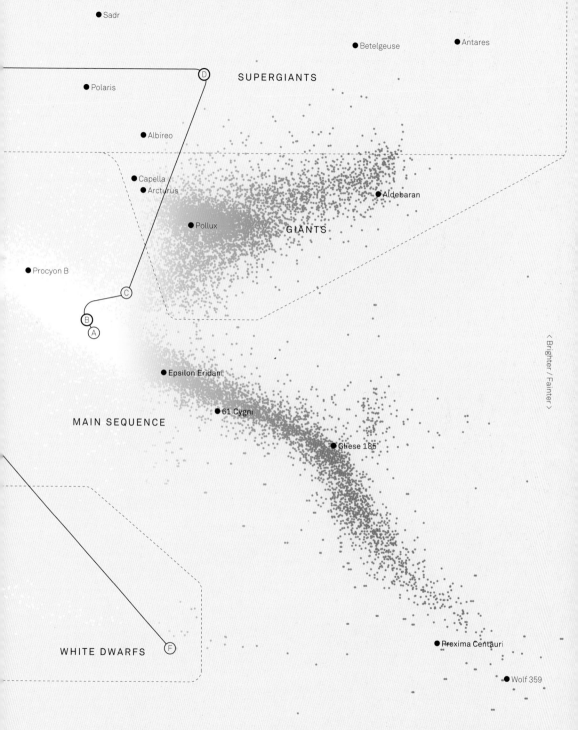

Sadr

Betelgeuse　　Antares

SUPERGIANTS

D

Polaris

Albireo

Capella
Arcturus

Aldebaran

Pollux

GIANTS

Procyon B

C

B
A

Epsilon Eridani

61 Cygni

MAIN SEQUENCE

Gliese 185

⟨ Brighter / Fainter ⟩

Proxima Centauri

WHITE DWARFS

F

Wolf 359

⟨ Hotter / Cooler ⟩

Van Biesbroeck's star

Life cycle of stars

The life of a star is determined by how much mass it has when it is born. The more massive it is at birth, the more quickly it burns through its fuel supplies. Big stars live fast and die young. Through nuclear fusion, stars convert their huge supplies of hydrogen into progressively heavier elements. It is actually the radiation and energy from these nuclear reactions that keeps a star from collapsing under its own gravity. Once the fuel is depleted the star can no longer support itself and explodes off its outer layers while the inner parts collapse. This marks the death of the star. The end product will depend on the star's mass when it dies.

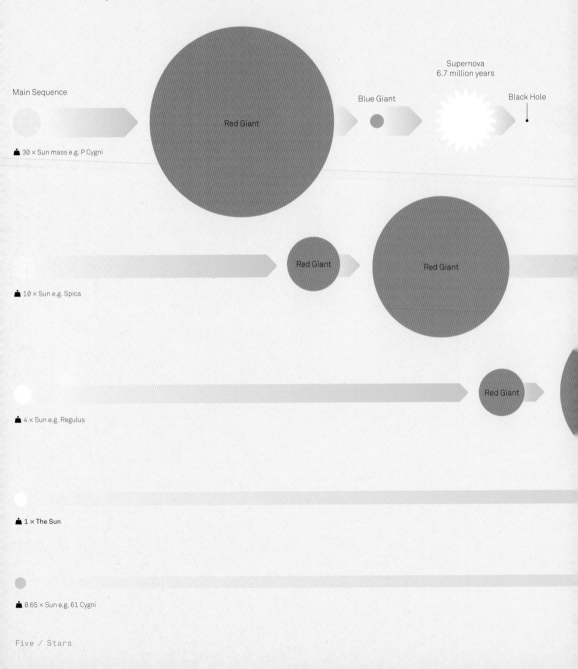

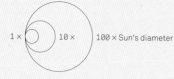

$1 \times$ $10 \times$ $100 \times$ Sun's diameter

Stars more massive than around 25 times
the Sun's mass collapse too rapidly to form
a neutron star, instead they form a black hole.

When stars more than eight times the mass
of the Sun end their lives in a supernova,
their inner layers collapse to form a neutron
star, which has a mass similar to the Sun
but a diameter of only around 20 km.

Supernova
27.6 M yr

Neutron Star

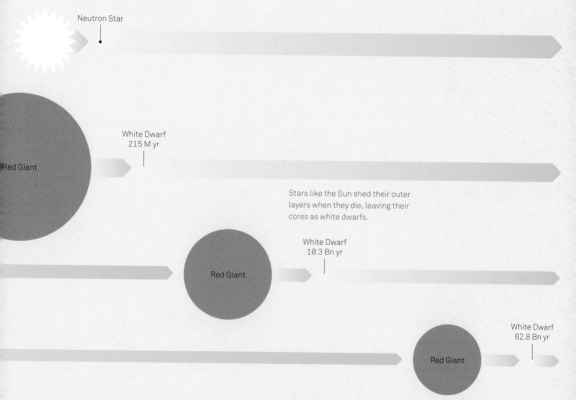

White Dwarf
215 M yr

Red Giant

Stars like the Sun shed their outer
layers when they die, leaving their
cores as white dwarfs.

White Dwarf
10.3 Bn yr

Red Giant

White Dwarf
62.8 Bn yr

Red Giant

Supernovae

When stars much more massive than our Sun reach the end of their lives they catastrophically explode as supernovae. These explosions can, briefly, outshine an entire galaxy. We estimate that about two or three occur each century in a galaxy the size of the Milky Way.

All supernovae we've found in the past 130 years have occurred in external galaxies and almost all were too faint to be seen by the unaided eye. The brightest supernova in living memory is SN 1987A which was seen in 1987 in the Large Magellanic Cloud.

These days we group supernovae into four main types – Ia, Ib, Ic and II – based on the elements observed in the spectra of their light. Although there are four types, there are two main causes.

Type Ia supernovae occur when a companion star spills matter onto a white dwarf until it eventually reaches a critical mass and explodes.

Type Ib, Ic and II supernovae occur when the nuclear fusion in very massive stars is no longer able to counteract gravity

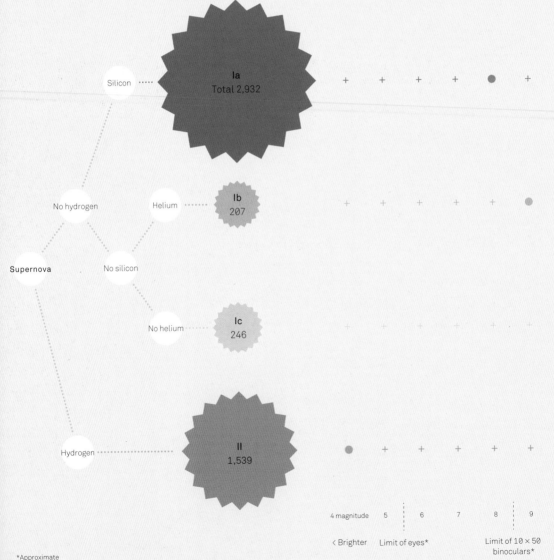

*Approximate

and the core collapses into either a neutron star or black hole. The type Ib and Ic are from stars that have lost most of their outer hydrogen layers through stellar winds in the final stages of their lives.

Over the past 130 years we've found more Type Ia supernovae than any other type and they make up the majority of the faintest supernovae. This is partly because these tend to be the brightest supernovae so we can detect them out to further distances.

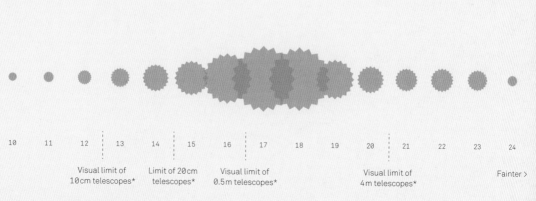

| 10 | 11 | 12 | 13 | 14 | 15 | 16 | 17 | 18 | 19 | 20 | 21 | 22 | 23 | 24 |

Visual limit of 10cm telescopes*

Limit of 20cm telescopes*

Visual limit of 0.5m telescopes*

Visual limit of 4m telescopes*

Fainter >

Pulsars

In 1967 Cambridge University student Jocelyn Bell noticed 'a bit of scruff' in the data from her radio telescope. It was a regularly pulsating burst of radio waves and she realised it was extra-terrestrial in origin. She named it *'LGM 1'* – short for *'Little Green Man 1'* – and *LGM 2 & 3* were found a little later. After discussion with others she realised she had discovered a type of dead star that had been predicted but never seen.

A star much more massive than the Sun ends its life as a supernova. While the outer parts explode, the inner parts collapse to a tiny remnant with a mass slightly more than the Sun compressed to the size of a city. During the collapse the star spins faster, the magnetic field gets concentrated, and the density becomes so high that protons and electrons are forced together into neutrons. The result is a neutron star and the surface can be spinning as fast as 15% of the speed of light.

Often, neutron stars emit beams of radio waves from each magnetic pole. If the rotation and magnetic axes differ, the beams sweep around the sky just like the beams from a

× Pulsar
◎ Pulsar has one or more companions
⊗ Pulsar associated with a supernova remnant

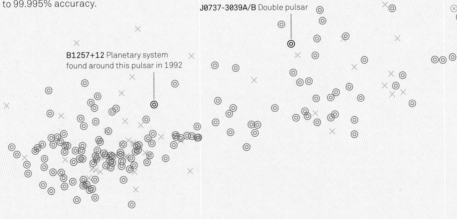

Vela Remnant of supernova which exploded over 10,000 years ago

Crab Supernova exploded in 1054 and recorded by Chinese astronomers. Pulsar discovered in 1968

Hitting the brakes >

Crab

One cubic centimetre of the Crab pulsar's magnetic field is equivalent to the output of a nuclear power plant. A cubic meter contains more energy than the total output of humanity.

J0737-3039A/B

In 2004 two pulsars were found orbiting each other. By carefully monitoring changes in their behaviour, astronomers have been able to test General Relativity to 99.995% accuracy.

J0737-3039A/B Double pulsar

B1257+12 Planetary system found around this pulsar in 1992

< Gentle

0.001 s / rotation   0.01 s / rotation   0.1 s / rotation

lighthouse. If the beams sweep past the Earth we see a flash of radio waves each time and we call it a pulsating neutron star or pulsar.

Comparing the rotation time with how fast a pulsar is slowing produces interesting results. Younger pulsars are found towards the top left and pulsars with companions are found in the lower left. As a pulsar ages it slows and the rate of slow-down reduces too; the pulsar moves down and to the right. At some point it crosses the 'pulsar death line' where the beams seem to switch off.

B1919+21 First pulsar discovered, 1967. Initially known as 'LGM 1'

Pulsar death line

× **J2144-3933** Very slow pulsar that challenges our understanding of the pulsar death line

1 s / rotation

10 s / rotation

We are stardust

Life is complex, relying on chemical reactions between a huge variety of atoms and molecules. The basis for all life on Earth is DNA, which is made of hydrogen, carbon, nitrogen, oxygen and phosphorus. But where do all these chemical elements come from? Most of the heavier elements are formed during supernovae, including some of the elements that are crucial to life as we know it. This material goes into the formation of new stars and planets.

In the beginning

In the beginning*, the Universe was a sea of the two lightest elements, hydrogen and helium, with very small amounts of lithium, boron and beryllium formed from subsequent reactions.

* Not quite the beginning; the first stable atoms didn't form until about 380,000 years after the Big Bang.

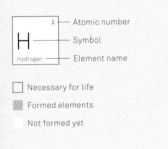

Atomic number
Symbol
Element name

☐ Necessary for life
▨ Formed elements
☐ Not formed yet

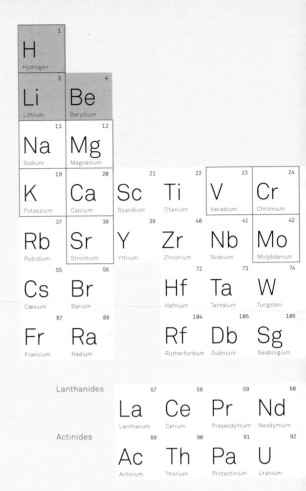

Lanthanides

Actinides

The hearts of stars
Stars like the Sun can form carbon, nitrogen, oxygen, neon, and silicon towards the end of their lives.

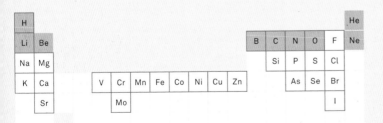

Stellar monsters
The energetic processes within the cores of massive stars produce elements that fill in half of the rest of the periodic table. Aluminium, silicon, and oxygen are the three most common elements found in the Earth's crust.

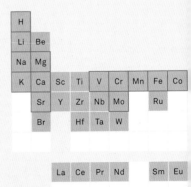

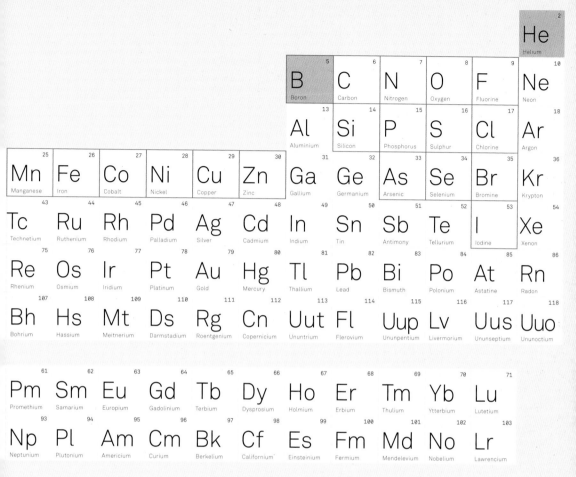

Ashes to ashes

Most of the heavier elements are formed during supernovae, the final death throws of the most massive stars. The material created in these explosions includes some of the elements that are crucial to life as we know it.

Precession

The north end of the Earth's axis of rotation points towards Polaris (or a point close to it), which is sometimes called the North or Pole Star. There's nothing particularly special about Polaris, though. This alignment of the North Celestial Pole with Polaris is just a coincidence, and hasn't always been the case. The Earth's axis is tilted by about 23.5 degrees relative to its orbit around the Sun. Over a period of tens of thousands of years this axis wobbles, changing position in the sky above the poles.

North Celestial Pole

When the pyramids were being built, the North Celestial Pole was close to the star Thuban, in Draco, and in 14,000 years hence the very bright star Vega will be the pole star.

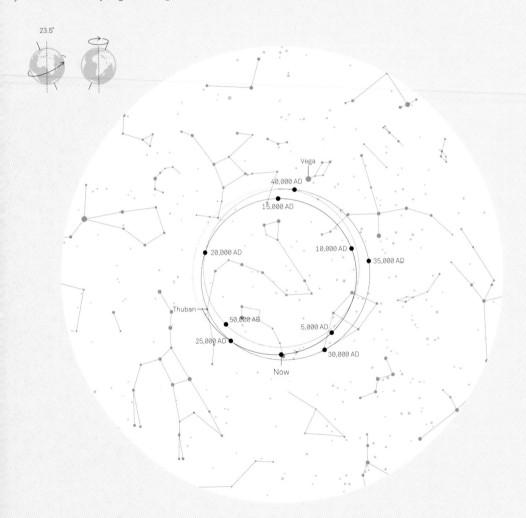

South Celestial Pole

There isn't currently a bright star near the South
Celestial Pole, but it is pointed to by the Southern Cross.
In a thousand years that won't be the case, and in 14,000
years the pole will be about ten degrees from the second
brightest star in the sky, Canopus.

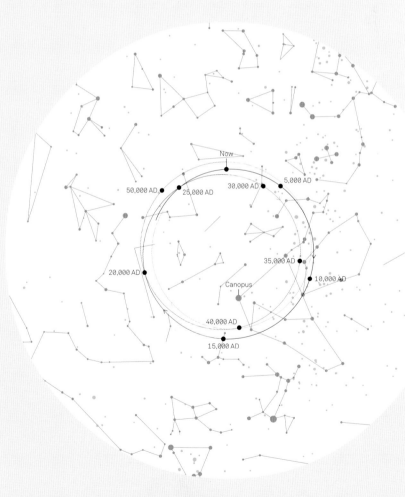

Six / Galaxies

The Milky Way

From the outside, our galaxy – the Milky Way – would look like two fried eggs placed back-to-back. We are in the white of the egg, roughly two-thirds of the way from the centre, so we see a band circling us. We can lay out the sky, as you would a globe, with the band of the Milky Way as the 'equator' line across the middle of the page and the centre of our galaxy at the centre of the map.

The left and right edges of the page join up. Looking at the 'equator' line means looking into the disk so we mostly see the local stars, dust and nebulae. Moving up or down from the 'equator' is to look up or down out of the disk giving a less obstructed view to the things beyond.

As well as stars, many other types of objects are visible in the sky, from clusters of stars which formed together, to galaxies both near and far.

Mrk 841

PSR J1836+5925

M 14

◇ NGC 6633

Cygnus A

NGC 7822

M 52

✕ Cygnus X-1

Trifid Nebula

◇ M 38

IC 1848   ◎ NGC 896

◇

Cygnus Superbubble ◎

M 11 ◇

Tadpole Nebula

Cassiopeja A ⊗

M 27

M 22 ◆

◎ NGC 281

Veil Nebula

California Nebula

Perseus Cloud

● Andromeda Galaxy

◇ The Pleiades

Andromeda Galaxy

A similar galaxy to our own Milky Way. At a distance of 2.5 million light years, it is the most distant object visible to the naked eye.

The Pleiades

An 'open cluster' of a few hundred stars. They formed together from the same cloud of gas and dust around 100 million years ago, and will gradually disperse over time.

● 3C 454.3

Legend

× Star
✳ Group of stars
◇ Open cluster
◆ Globular cluster

◎ Nebula
⊗ Supernova remnant
⬤ Galaxy
◔ Spiral galaxy

↩ Barred galaxy
◖ Lenticular galaxy
◣ Elliptical galaxy

North Celestial Pole
Galactic Centre
South Celestial Pole

◣ M 87

⬤ 3C 273

⬤ 3C 279

3C 273

In the 1950s radio astronomers found hundreds of very bright objects all over the sky. Many of these were identified as distant galaxies with supermassive black holes at their centres. These objects are now called quasars.

Centaurus A

Looks like an unassuming galaxy in visible light but has two huge lobes spurting out in opposite directions from the supermassive black hole at its heart.

◎ QSO J1512-0906

Vela Supernova Remnant

The remains of a massive star that exploded over 11,000 years ago in the direction of the constellation Vela. The centre contains a neutron star.

✳ Upper Scorpius
✳ Rho Ophiuchi

⬤ Centaurus A
◆ Omega Centauri

SN 437 ×

◇ NGC 4755

◎ Carina Nebula

◇ IC 2602

◇ M 93

⊗ Vela Supernova Remnant

Rosette Nebula ◎

Crab Supernova Remnant ◔

◇ M 41

Lambda Orionis ✳

◎ Flame Nebula
◎ Orion Nebula

Omega Centauri

The largest globular cluster in our galaxy. This roughly spherical ball of stars formed together several billion years ago, and they are bound together by gravity.

Orion Nebula

Stars form from clouds of gas and dust called nebula. The brightest of these, as seen from Earth, is the Orion Nebula which lies around 1,500 light years away.

↩ Large Magellanic Cloud

⬤ Small Magellanic Cloud

Large and Small Magellanic Clouds

Small galaxies which are orbiting our own Milky Way. They are visible to the naked eye from the southern hemisphere.

× SN 2006dd

The invisible galaxy

What we can see with our eyes is only part of the picture.
Our first glimpses beyond the visible were with radio
telescopes. These showed some of the same familiar
features such as the band of the Milky Way but also
strange new objects. With space telescopes we can now
see across almost the entire electromagnetic spectrum.

1 / Gamma-ray

The Fermi Gamma-ray Space Telescope allows us to study subatomic particles at energies much greater than in ground-based particle accelerators. It lets us see the effects of black holes and other highly energetic events in the Universe.

2 / Infrared

The Infrared Astronomical Satellite (IRAS) was a joint satellite of the US, UK and the Netherlands. Infrared light is particularly useful in detecting wisps of warm dust called galactic cirrus.

3 / Microwave

The Planck satellite was a European Space Agency mission launched in 2009. It shows gas and dust in our galaxy but also, high above and below the plane, sees the earliest light in the Universe that set off just 380,000 years after the big bang.

4 / X-ray

ROSAT was a joint German, US and UK X-ray satellite. X-rays are produced by matter heated to millions of degrees. They can come from cosmic explosions and high-speed matter. The black stripes are not real features; they are patches missed due to problems on the satellite.

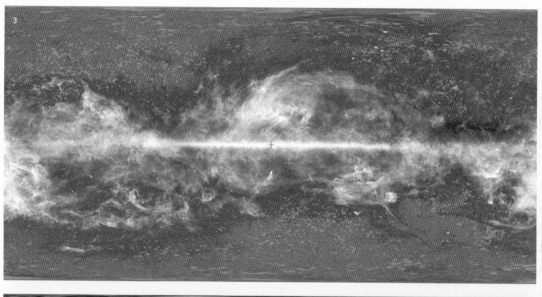

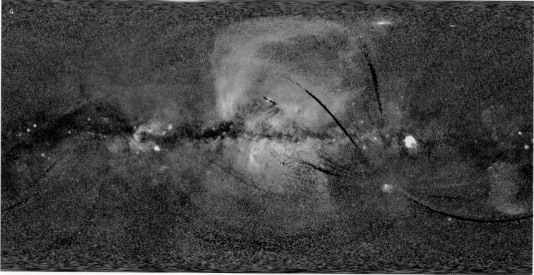

Polarisation of the Galaxy

The Milky Way has a magnetic field generated by the motion of charged particles. While we can't see this magnetic field directly, we can see its side-effects, such as the alignment of tiny grains of cosmic dust. ESA's Planck satellite showed this pattern of alignment over the entire sky, revealing intricate structures both within the Milky Way and in nearby galaxies. The patterns highlight some of the nearby regions where stars and dust are forming, creating distortions and turbulence in the field.

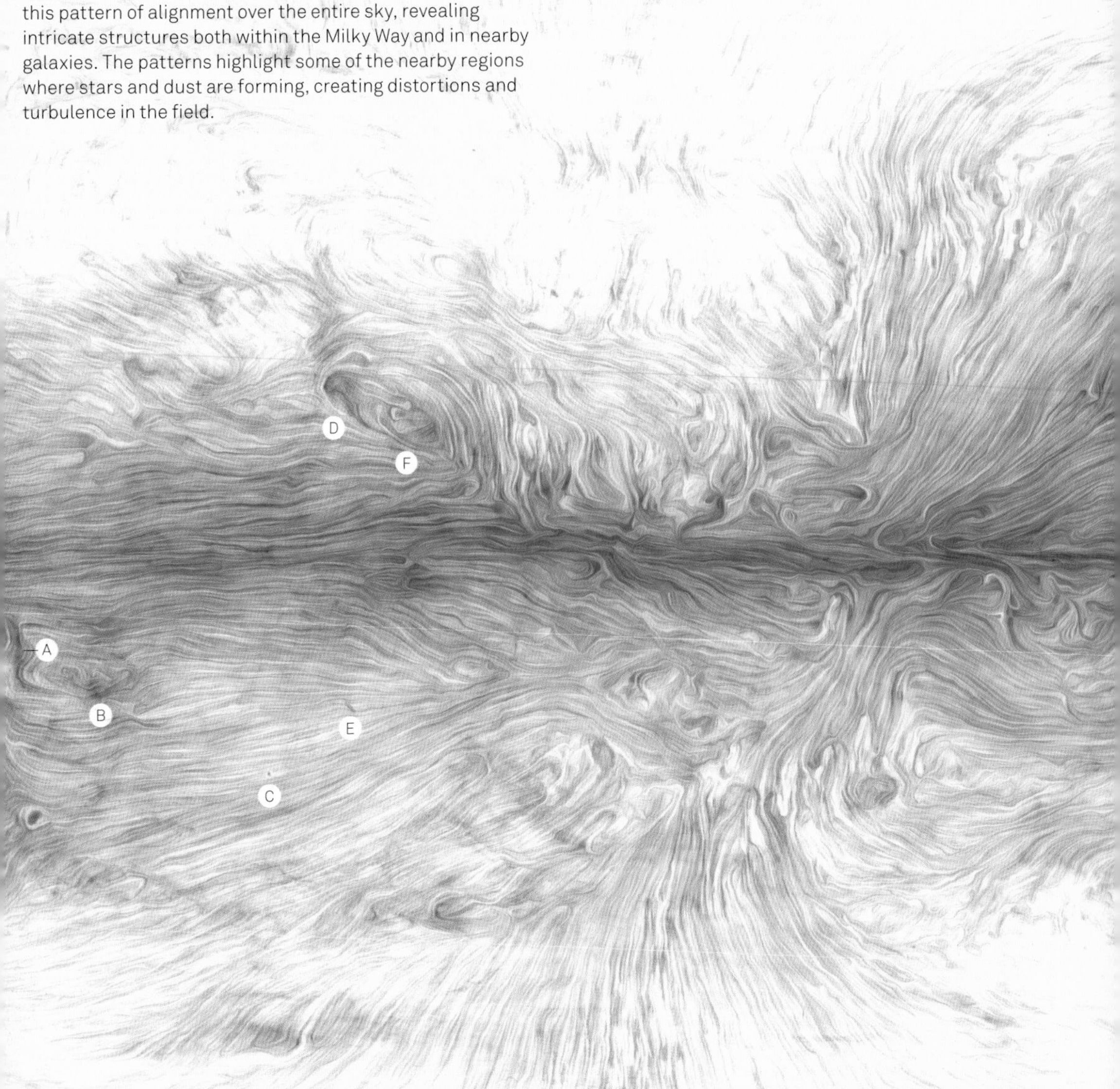

A / Taurus molecular cloud
B / Perseus molecular cloud
C / Triangulum Galaxy

D / Polaris flare
E / Andromeda Galaxy
F / Cepheus flare

G / Rho Ophiuchi
H / Small Magellanic Cloud
I / Carina Nebula

J / Large Magellanic Cloud
K / Vela molecular ridge
L / Orion molecular cloud

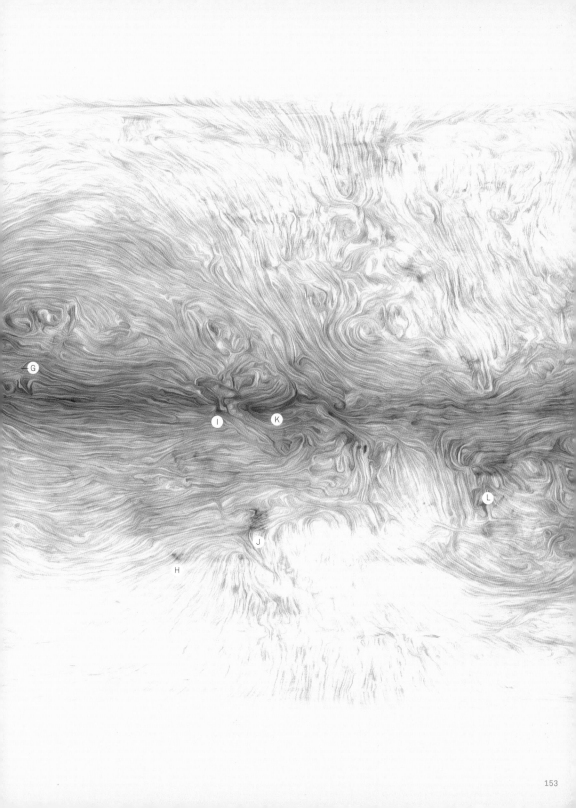

Structure of the Milky Way

Most of the stars we see are within a few thousand light years of the Sun, and so relatively local on the scale of the Galaxy. Early observations showed that our galaxy is a disc shape, around 100,000 light years wide and just a few thousand light years thick. Observations in the 1980s showed that some older stars are in a thicker disk around 30,000 light years thick. Around the disk is a roughly spherical 'halo' of stars. These stars are amongst the oldest in the galaxy, and the halo contains a number of ancient globular clusters.

By mapping the galaxy at radio and infrared wavelengths, it is possible to see through much of the obscuring dust and build a map of the 3D structure. We now know that there are two main spiral arms that extend from the ends of a 30,000-light-year-long bar in the centre. These spiral arms aren't fixed structures, but show where there is a higher density of stars. They move independently of the stars, much as a traffic jam moves backwards along a motorway while the cars move forwards.

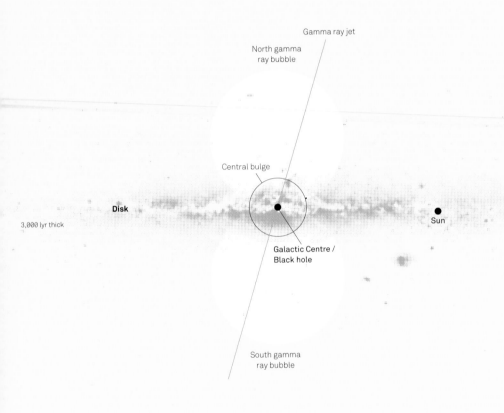

Gamma ray jet

North gamma
ray bubble

Central bulge

Disk

3,000 lyr thick

Sun

Galactic Centre /
Black hole

South gamma
ray bubble

In 2010, NASA's Fermi satellite found evidence that there are bubbles of hot gas blowing from the centre of the Milky Way. These may be from the explosions of massive stars, or possibly linked to the central supermassive black hole in the Galactic Centre.

100,000 light years across

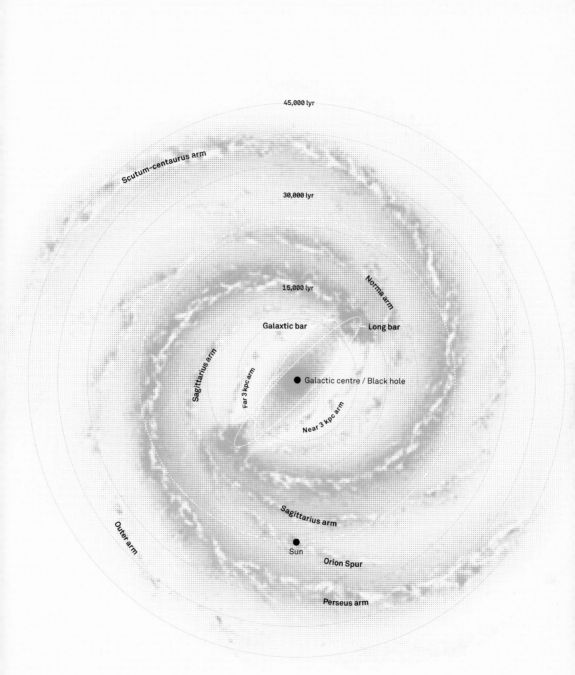

45,000 lyr

Scutum-centaurus arm

30,000 lyr

Norma arm

15,000 lyr

Galaxtic bar

Long bar

Sagittarius arm

Far 3 kpc arm

● Galactic centre / Black hole

Near 3 kpc arm

Outer arm

Sagittarius arm

● Sun

Orion Spur

Perseus arm

The Sun is located around 28,000 light years from the centre in the 'Orion Spur', sat between the two major spiral arms, and just below the centre of the disk.

The local sheet of galaxies

The Milky Way is just one of many galaxies. The closest galaxies to us are the dwarf irregular galaxies of the Large & Small Magellanic Clouds and the Sagittarius Dwarf Galaxy. The nearest large galaxy is the Andromeda Galaxy which is about 2.5 million light years away.

The Local Group consists of the Milky Way, Andromeda, and around 50 dwarf galaxies within about five million light years. Beyond the Local Group, within about 25 million light years, there are another 40 – 50 large, bright galaxies. Many of these form the Local Sheet; a thin, pancake-like cluster of galaxies inclined at eight degrees to the Local Supercluster (100,000 galaxies in a region 500 million light years across).

In the direction of Leo there is a small group of galaxies known as the *M96 Group*. This is physically separate from the Local Group but is part of the same Local Supercluster.

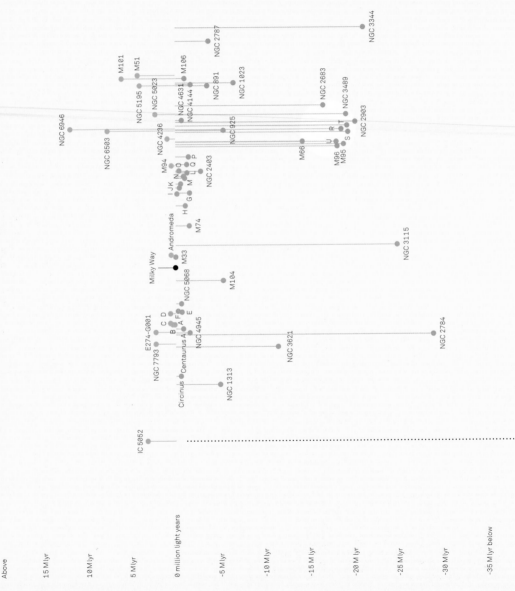

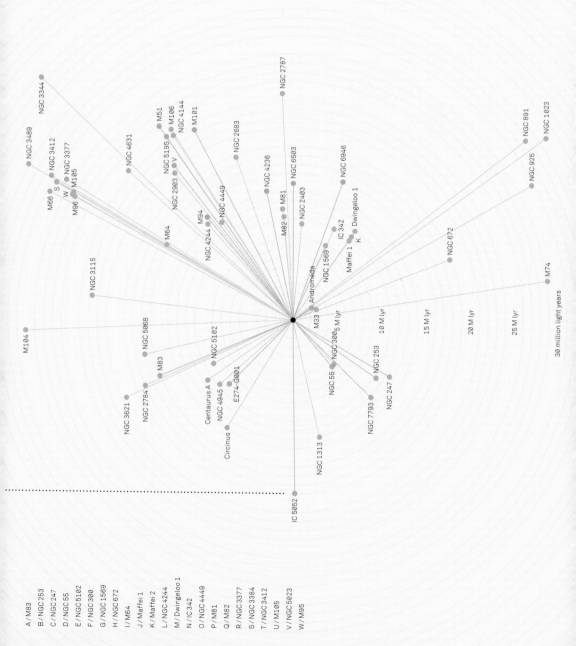

NGC 3344

NGC 2787

NGC 3489
NGC 3412
S · NGC 3377
M51
M106
NGC 4144
NGC 4631
W · NGC 3377
M101
M105
M66
S
M96
NGC 5195
V
NGC 2903
NGC 4449
NGC 2683
NGC 4236
NGC 6503
NGC 6946
NGC 891
M64
M94
NGC 4244
M82 · M81
NGC 2403
NGC 925
NGC 1023
IC 342
Dwingeloo 1
Maffei 1
K
NGC 672
M74
NGC 3115
Andromeda
NGC 1569
M104
NGC 5068
NGC 5102
M33
NGC 55
NGC 300 5 M lyr
10 M lyr
15 M lyr
20 M lyr
25 M lyr
M83
NGC 253
30 million light years
NGC 3621
NGC 2784
Centaurus A
E274-G001
NGC 247
NGC 4945
NGC 7793
Circinus
NGC 1313

IC 5052

The 30,000 nearest galaxies

When we look out into the Universe we see that galaxies are not evenly spread. When they formed, from the primordial soup of gas, the force of gravity started to pull them together. They can be found inhabiting huge clusters and are strung out in long streams across the sky. Our own galaxy is being pulled toward a nearby cluster, which is itself being slowly drawn towards a much more distant supercluster dubbed *The Great Attractor*.

Zone of avoidance (plane of the Milky Way)

Galaxy type
🌀 Spiral
◯ Elliptical
● Other

Distance
● ● ● ● near – far

A / Perseus-Pisces Supercluster
B / Shapley Supercluster
C / Virgo Supercluster
D / The Great Attractor
E / Coma Supercluster

The galactic zoo

In 1926 astronomer Edwin Hubble proposed a method to classify the shapes of galaxies. This has become known as the Hubble Sequence or, more commonly, the Hubble Tuning Fork. It has ball-shaped galaxies on one end and spiral galaxies on the other. The fork-shape arises due to splitting spiral galaxies into groups with or without bars in their centres.

Classifying galaxies based on what they look like takes time and is a surprisingly difficult task for computers.

In 2007, with hundreds of thousands of galaxies found in the Sloan Digital Sky Survey, astronomers had to find a new way to process the data. They launched a website named galaxyzoo.org that asked members of the public to classify galaxies. The site turned out to be incredibly successful and, by 2010, nearly 84,000 members of the public had provided over 16 million detailed classifications of over 300,000 galaxies. Between them they compiled the largest and most reliable database of galaxy shapes yet.

◯ Number of galaxies

Ellipticals

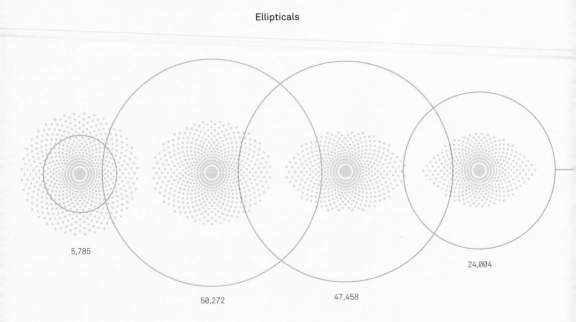

5,785

50,272

47,458

24,004

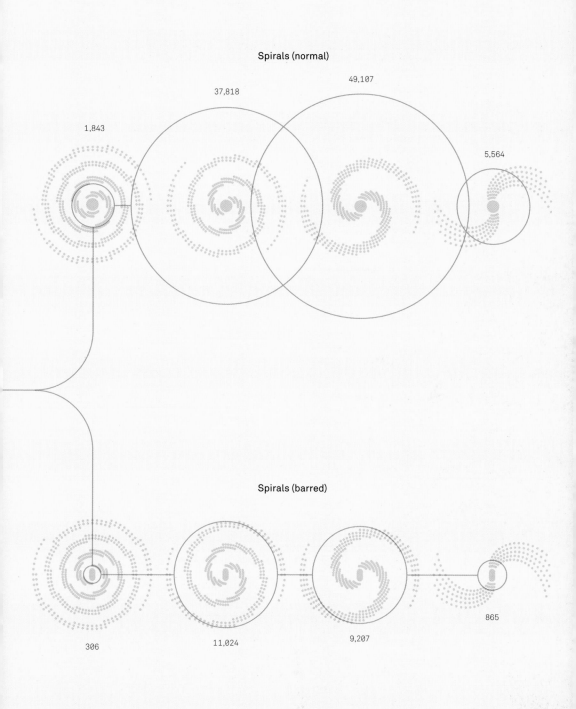

Spirals (normal)

37,818

49,107

1,843

5,564

Spirals (barred)

306

11,024

9,207

865

Seven / Cosmology

Empty space

Space is empty. Really empty. The average density
is one hydrogen atom per cubic metre.

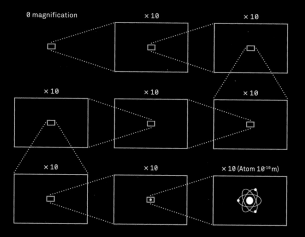

Models of the Universe

The Universe is everything there is. Our understanding of what that means, and our place in it, has changed considerably over the past few thousand years.

Many, but not all, early ideas put the Earth at the centre. Around the Earth were the planets on rotating spheres and beyond those the 'fixed' stars. Some argued the stars were distant versions of our own Sun with their own planetary systems.

The Platonic universe had the Earth at the centre with the planets orbiting in rotating spheres. It was clear though that this didn't explain the observed motions of the planets and so Ptolemy added eccentric circles and epicycles: the planets orbited points which, in turn, orbited a point offset from the Earth.

In medieval times, models of the Universe were still influenced by Plato but included current religious teachings and even novel shapes such as Hildegaard of Bingen's cosmic egg.

By the 16th Century mathematics and observation started to play a larger role. Copernicus found that

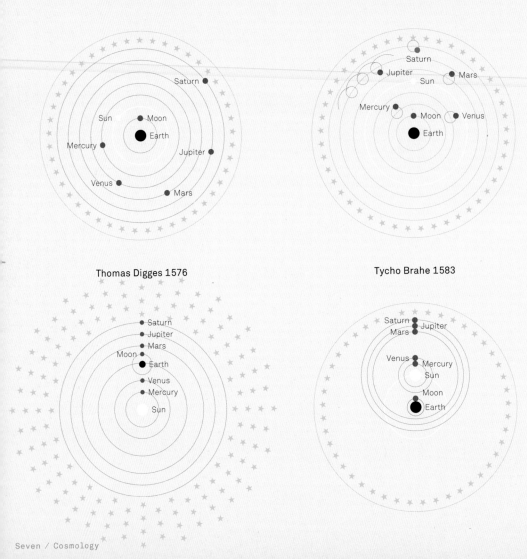

Plato 427-347 BCE

Ptolemy 2nd Century CE

Thomas Digges 1576

Tycho Brahe 1583

calculations were easier if the Sun was placed at the centre with the planets on perfect, circular orbits.

Thomas Digges modified this view by dispersing the stars evenly throughout a much larger universe. Danish astronomer Tycho Brahe used his observations to promote an Earth-centred universe but with a compromise: he had the Sun orbiting the Earth and all the other planets orbiting the Sun. Kepler brought back Copernicus's Sun-centred universe but with the planets orbiting on ellipses.

As the Enlightenment continued, observations and theory advanced. Thomas Wright and Immanuel Kant argued that the existence of the band of the Milky Way meant that stars must be distributed in some kind of disk around us. It was even suggested that our Milky Way was one of 'many islands' in the Universe.

Today our observations are far superior to those of our ancestors but, as has often been the case, they hint that our current model isn't quite the whole picture.

Hildegaard of Bingen 1142

Nicolaus Copernicus 1543

Thomas Wright 1750

The cosmological distance ladder

Measuring distances in the Universe isn't easy when you are confined to planet Earth. We can measure the distances to nearby objects using simple geometry but as we look out further into the Universe we have to use different methods. Many methods rely on finding a 'standard candle' – something for which we know the intrinsic brightness – and then finding its distance from its apparent brightness. Each method works over different ranges of distance so if we find enough overlapping methods we can connect them together to create what is called the cosmological distance ladder. ESA's Gaia satellite is currently measuring parallaxes out to much greater distances.

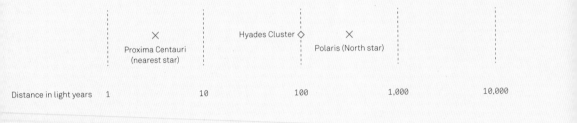

		Hyades Cluster ◇	✕		
	✕		Polaris (North star)		
	Proxima Centauri (nearest star)				

Distance in light years   1   10   100   1,000   10,000

Cepheid variable

Main sequence fitting

Spectroscopic parallax

Parallax

Parallax (Gaia)

Radar

Radar

For nearby objects within our Solar System we can use radar to directly measure the distance. Radar measurements of the planet Venus helped define the scale of the Earth's orbit.

Parallax

If you hold up a finger in front of you and look at it with one eye at a time you see that it appears to move compared to distant objects. We can use exactly the same technique by measuring a star's position from different sides of the Earth's orbit. A bigger change in position means the star is closer.

Spectroscopic parallax

If a star is bright enough it is possible to measure its spectrum of light. Identifying dark absorption lines in this spectrum lets you work out what the intrinsic brightness of the star should be and from that work out the distance.

Main sequence fitting

We measure the apparent brightness and colour of all stars in a cluster which we assume formed at the same time and the same distance from Earth. Plotting these values on an H-R diagram we can see a group called the 'main sequence'.
By calculating how much fainter or brighter this 'main sequence' is than that of another cluster, we can find the relative distances.

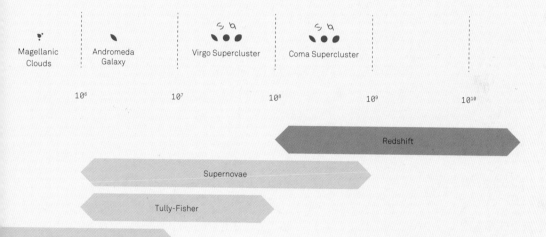

Magellanic
Clouds

Andromeda
Galaxy

Virgo Supercluster

Coma Supercluster

10^6 10^7 10^8 10^9 10^{10}

Redshift

Supernovae

Tully-Fisher

Cepheid variables

These variable stars brighten and fade as they expand and contract. The time it takes to pulse is directly linked to the star's true brightness. Measuring the period the true brightness can be found and comparing this to the apparent brightness gives us the distance.

Tully-Fisher

In 1977, astronomers Brent Tully and J.R. Fisher noticed that the rotation velocity of spiral galaxies was related to their intrinsic brightness. By measuring the velocity through the Doppler Effect, it is then possible to determine the distance.

Supernovae

A Type Ia supernova occurs when a white dwarf reaches a critical mass (1.44 solar masses) and explodes. As the explosion occurs at a set mass, the true brightness of the explosion will always be the same. By measuring the apparent brightness and knowing the true brightness, the distance of the supernovae can be found.

Redshift

The Universe is expanding and that means the light from distant galaxies is stretched (redshifted) as it travels towards us. How much the light is stretched depends on how far away the galaxy is from us and this can be found by measuring its spectrum.

The cosmic web

Measuring distances to galaxies takes a lot of time, particularly when those galaxies are incredibly faint and distant. The Sloan Digital Sky Survey uses a dedicated telescope to do just that, and measures the distances to millions of galaxies, reaching out billions of light years. By looking at a narrow slice through space, away from contamination from our own Milky Way, the cosmic web of galaxy clusters becomes apparent, even when only showing the brightest galaxies.

Galaxies are seen to lie along long filaments, stretching billions of light years. Between the filaments are huge voids, with very few galaxies. This pattern continues out to huge distances, though the galaxies get harder to see.

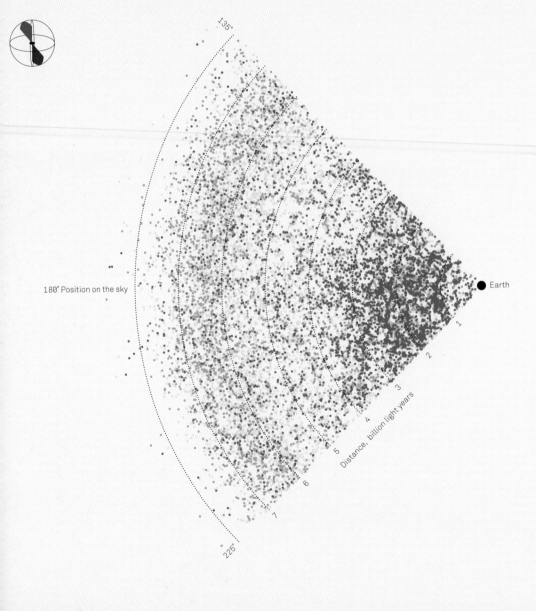

135°

180° Position on the sky

225°

Earth

1

2

3

4

5

6

7

Distance, billion light years

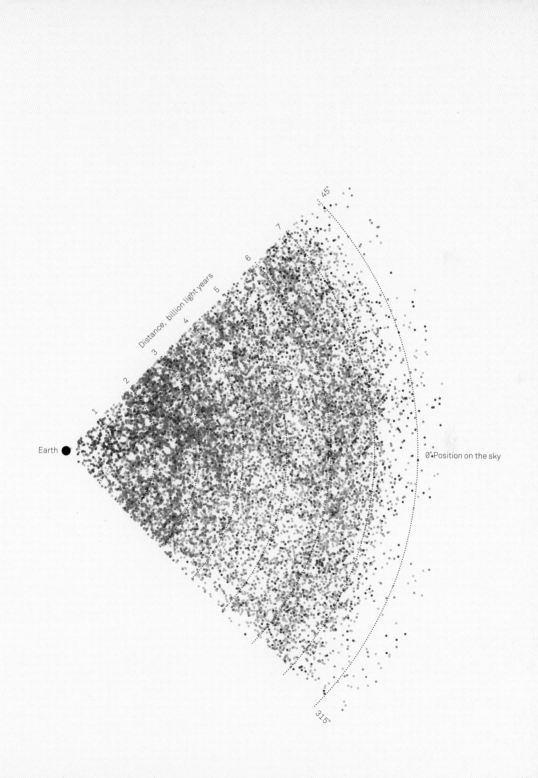

Earth ●

Distance, billion light years

1 2 3 4 5 6 7

45°

0° Position on the sky

315°

What is the Universe made of?

When we look at the night sky we primarily see stars, but in fact they're only a tiny fraction of the Universe. In terms of mass they're outnumbered 10:1 by interstellar gas, dust and subatomic particles, which are largely invisible to the optical telescopes.

Even with all this, normal matter is outweighed 5:1 by 'dark matter'. This elusive component of the Universe doesn't emit, absorb or scatter light, but does have a gravitational pull.

That still accounts for only about one third of the Universe. The majority of the energy in the Universe is due to the truly mysterious 'dark energy', which acts to push apart clusters of galaxies and accelerate the expansion of the Universe as a whole.

- Dark energy 68.3%
- Dark matter 26.8%
- Normal matter 4.9%
  - Stars 0.5%
  - Gas 4%
  - Neutrinos 0.3%
  - Dust < 0.1%

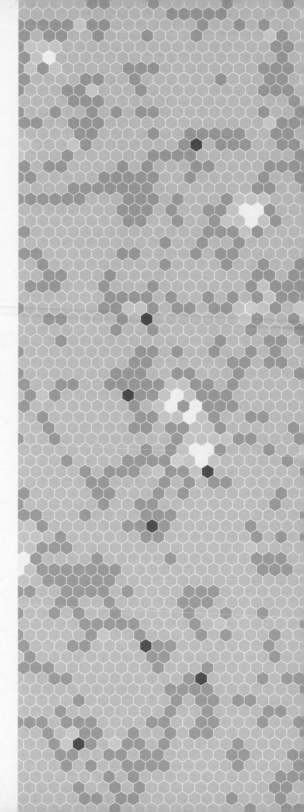

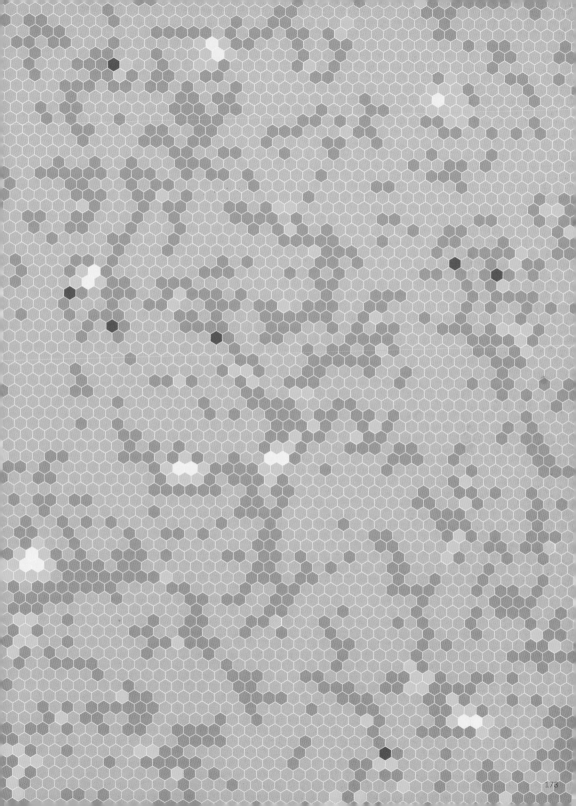

Evolution of the Universe

We can't describe the conditions right at the start of the Universe using current theories of physics, but we can get pretty close. Despite the huge temperatures involved, our current understanding allows us to understand the processes that took place within the first fraction of a second.

As the Universe expanded it cooled, and the constituents of the matter we're made of all formed within the first three minutes. It was too hot for atoms to exist until 380,000 years later, and the Universe was initially opaque to light. It is thought that the first stars formed a few hundred million years later, and that the galaxies built up over the next billion years. We think that the largest structures in the Universe – superclusters of galaxies – are due to tiny quantum fluctuations that originated in the first tiny fraction of a second after the Big Bang – the one period which we truly don't understand.

By the 1990s we thought we knew the future fate of the Universe. Then, in 1998, observations of distant supernovae showed that around 4 billion years ago something unexpected had happened. The expansion of space seems to have begun to accelerate, pushed apart by a mysterious 'dark energy'. The expansion seems to be real but nobody knows what this dark energy is.

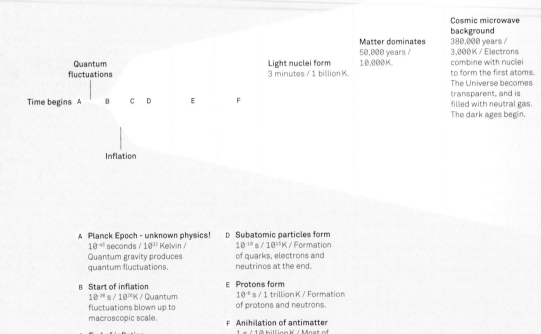

—— Radius of the visible universe

Quantum fluctuations

Time begins A   B   C   D   E   F

Inflation

Light nuclei form
3 minutes / 1 billion K.

Matter dominates
50,000 years /
10,000 K.

Cosmic microwave background
380,000 years /
3,000 K / Electrons combine with nuclei to form the first atoms. The Universe becomes transparent, and is filled with neutral gas. The dark ages begin.

A   Planck Epoch - unknown physics!
10^{-43} seconds / 10^{32} Kelvin / Quantum gravity produces quantum fluctuations.

B   Start of inflation
10^{-36} s / 10^{28} K / Quantum fluctuations blown up to macroscopic scale.

C   End of inflation
10^{-32} s / 10^{27} K / Universe is almost completely uniform. Radiation dominates the Universe.

D   Subatomic particles form
10^{-10} s / 10^{15} K / Formation of quarks, electrons and neutrinos at the end.

E   Protons form
10^{-6} s / 1 trillion K / Formation of protons and neutrons.

F   Anihilation of antimatter
1 s / 10 billion K / Most of the matter in the Universe is dark matter.

Now
13.8 billion
years / 2.7 K.

**Dark energy
dominates**
10 billion years / 4 K /
The expansion starts
to accelerate.

Milky Way forms
5 billion years / 6 K.

Galaxies merge
3 billion years / 7 K /
Star formation is at
its peak.

First galaxies
1 billion years / 15 K
/ First galaxies grow
by merging together.

First stars
500 million years /
30 K / First stars
form, reionising
most of the gas
in the Universe.

Powers of ten

From the tiniest sub-atomic particles to the entire observable Universe, the scale of the Universe is truly vast.

As an attempt to get some understanding of the range let's start at the unbelievably tiny scale of a proton. We can zoom out by a factor of 10 and see the nucleus of an atom. Zooming out by a factor of 100,000 we get to the size of a water molecule and another factor of ten takes us to a DNA strand. Stepping up our zoom by a further 10,000 takes us to the width of a human hair and the scales we encounter every day.

Astronomy deals with enormous scales, often so vast as to be unimaginable. Even on local scales, the Moon is 380,000 kilometres away, which is a huge distance by human standards. The standard unit of length in astronomy is the light year, which is approximately 10 million million kilometres. Using these terms the scales are still huge – the Andromeda Galaxy, the Milky Way's nearest large neighbour, is over two million light years away.

The Universe seen on these large scales is just as empty as the Universe seen on the smallest scales. However you look at it, space really is empty!

B lyr / billion light years
M lyr / million light years
k lyr / thousand light years
T km / trillion kilometres
B km / billion kilometres
M km / million kilometres
km / kilometre
m / metre
cm / centimetre
mm / millimetre
μm / millionth of a metre
nm / billionth of a metre
pm / trillionth of a metre
fm / quadrillionth of a metre

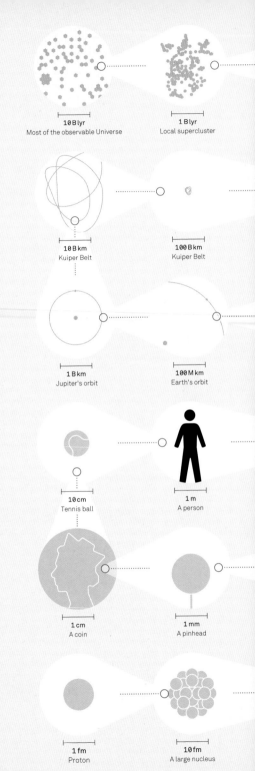

10 B lyr
Most of the observable Universe

1 B lyr
Local supercluster

10 B km
Kuiper Belt

100 B km
Kuiper Belt

1 B km
Jupiter's orbit

100 M km
Earth's orbit

10 cm
Tennis ball

1 m
A person

1 cm
A coin

1 mm
A pinhead

1 fm
Proton

10 fm
A large nucleus

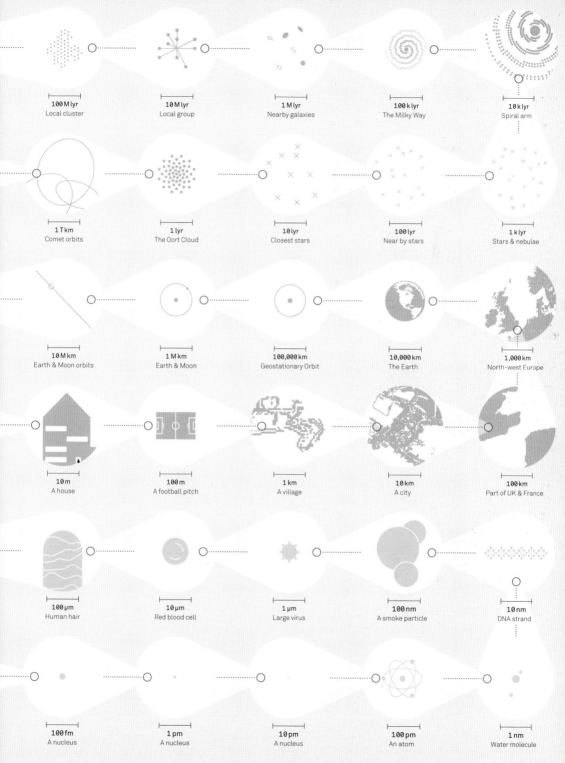

100 M lyr
Local cluster

10 M lyr
Local group

1 M lyr
Nearby galaxies

100 k lyr
The Milky Way

10 k lyr
Spiral arm

1 T km
Comet orbits

1 lyr
The Oort Cloud

10 lyr
Closest stars

100 lyr
Near by stars

1 k lyr
Stars & nebulae

10 M km
Earth & Moon orbits

1 M km
Earth & Moon

100,000 km
Geostationary Orbit

10,000 km
The Earth

1,000 km
North-west Europe

10 m
A house

100 m
A football pitch

1 km
A village

10 km
A city

100 km
Part of UK & France

100 µm
Human hair

10 µm
Red blood cell

1 µm
Large virus

100 nm
A smoke particle

10 nm
DNA strand

100 fm
A nucleus

1 pm
A nucleus

10 pm
A nucleus

100 pm
An atom

1 nm
Water molecule

Eight / Other worlds

Finding exoplanets

An exoplanet is one that orbits a star other than the Sun.
We've long suspected that other planetary systems existed
but it wasn't until the 1990s that techniques first let us confirm
them. How do you find one? There are several methods...

● Blips (microlensing)

There is a small chance that, from our
viewpoint on Earth, one star will happen
to pass in front of another. When this
happens, the gravity of the closer star
bends the light from the more distant star
around it and increases its brightness.
If the closer star has a planet, that
can also magnify the background star
causing an extra blip in the brightness.
This method can detect small planets
but, unfortunately, the blips are one-off
events that only last a few weeks so
follow-up can be difficult.

● Pictures (direct imaging)

Seeing planets around stars is hard
because a star is incredibly bright
compared to a planet. It is like trying
to see a fly next to a stadium floodlight.
To give ourselves more of a chance we can
change to infrared light where the planet
will be brighter. For example, the Sun is
about a billion times brighter than Jupiter
in visible light but only 100 times brighter
in infrared. To help even more we can try
to block the light from the parent star.
It is easier to find warmer planets further
from their star with this method.

● Side-to-side (astrometry)

We usually think of a planet orbiting a
star but things are actually more subtle.
Both the star and planet orbit their
common centre of mass. If you make
very accurate measurements of the
position of the star over time, you can
see it move around its orbit. The effect
is larger when the planets are in larger
orbits. This method is incredibly difficult
because the movement of the star is so
tiny. Nevertheless, the European Space
Agency's Gaia spacecraft is expected to
find many planets this way.

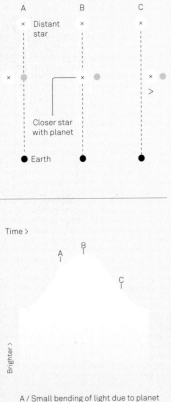

A / Small bending of light due to planet
B / Small bending of light due to star

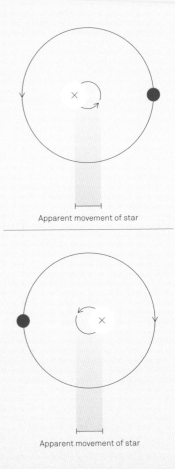

Apparent movement of star

Apparent movement of star

● Ticks (pulsar timing)

The first confirmed exoplanets were
found around a type of star known as a
pulsar which emits beams of radiation
much like a lighthouse. The flashes from
a pulsar act as a very precise clock.
Extremely accurate measurements will
show changes in that clock due to the
pulsar orbiting a common centre of mass.
The planets found by this method were
surprising as they would not have been
expected to survive the explosion that
created the pulsar.

● Winks (transit)

If the plane of the orbit of a planet
happens to be aligned with the Earth,
the planet will pass between its parent
star and us. As it transits its star, it
blocks out a small amount of its light.
The amount of dimming depends on the
size of the planet; a larger planet blocks
more of the light. This dimming repeats
every time the planet orbits the star
so the time between each dip gives
the period of the orbit.

● Wobbles (radial velocity)

As the star orbits, it moves slightly
towards or away from us. As it moves
towards us, the Doppler Effect slightly
shifts its light to bluer colours. As it
moves away from us the light becomes
slightly redder. By measuring the
spectrum of the star this movement can
be measured and any planets inferred.
With this technique it is easier to spot
larger planets because they cause larger
wobbles. It is also easier to find planets
on smaller orbits because they cause
a faster wobble.

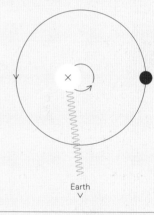

Earth
∨

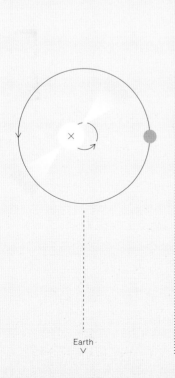

Earth
∨

Time >

Brighter >

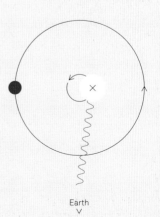

Earth
∨

Exoplanet discoveries

The first planet around a star other than the Sun was
detected in 1989. That detection was tentative and it
wasn't until 1992 that three more were found around
the unlikely body of a pulsating neutron star. In 1995,
51 Peg b was found, around a Sun-like star, and it opened
the door for many more in the years that followed.

In 2009, NASA's Kepler spacecraft launched with a mission
to find planets via the transit method: looking for planets
obscuring the light from their host stars as they orbit.
The mission has proved hugely successful having found
over half of the confirmed exoplanets currently known.
The mission continues to find planets and, as it is due
to operate well into 2016, should find many more.

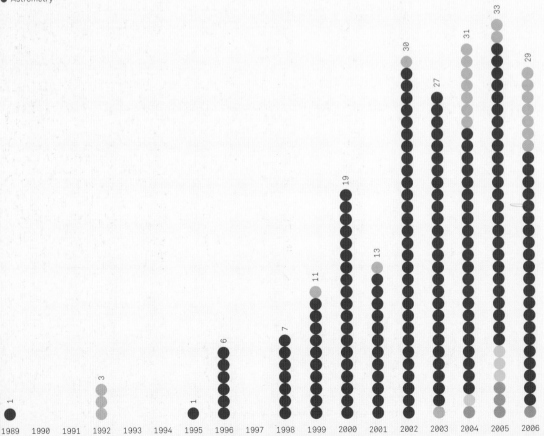

● Transit
● Radial velocity
● Pulsar timing
● Microlensing
● Direct imaging
● Astrometry

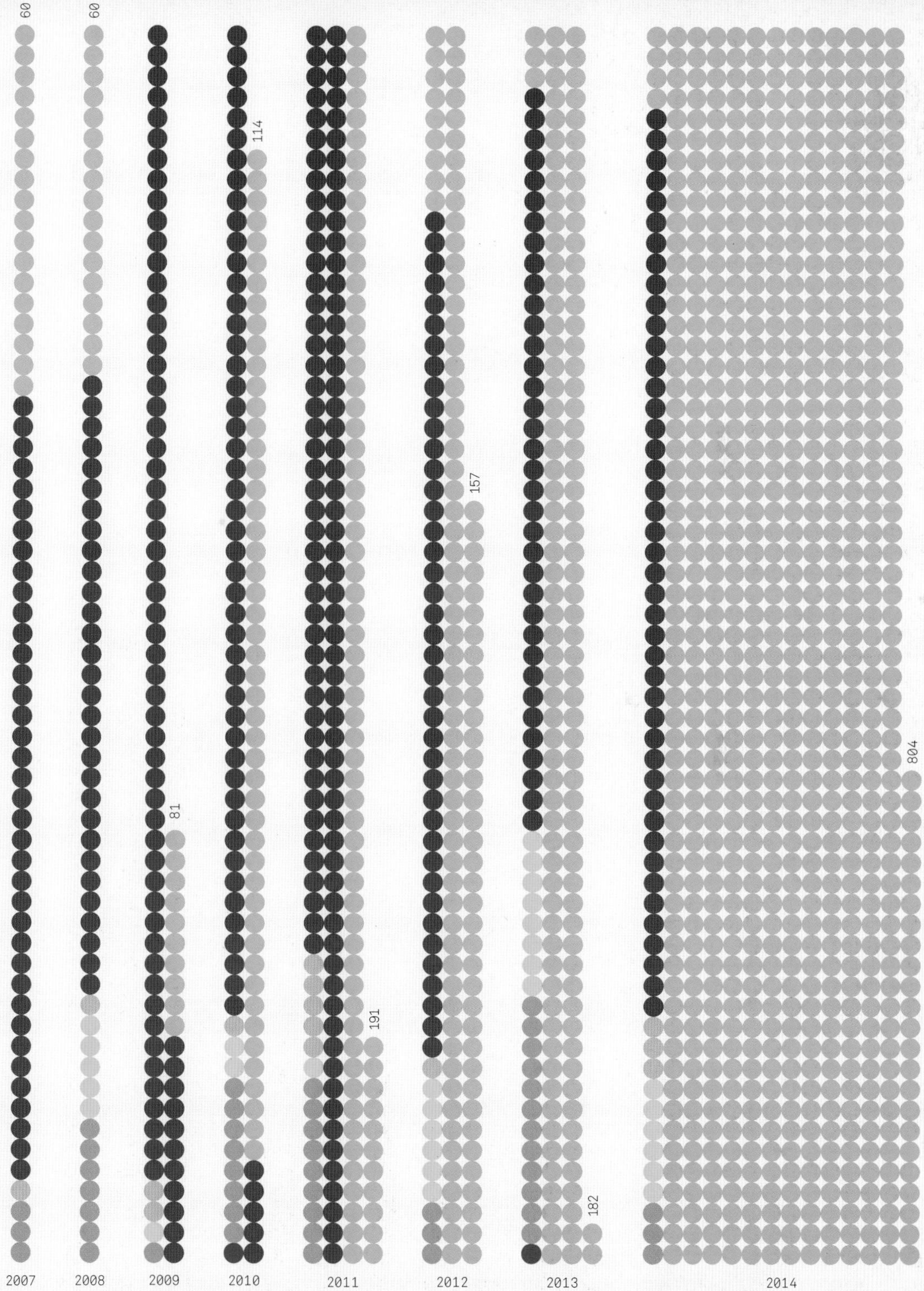

60

60

114

81

157

191

182

804

2007 2008 2009 2010 2011 2012 2013 2014

All known exoplanets

There are now over 1,800 confirmed planets orbiting other stars. The number is growing rapidly so will be out-of-date by the time you read this. If we place them all together, to scale, we get an idea what that looks like. The first thing to notice is the number of large planets. This is partly because there really are lots of planets bigger than Jupiter, but also because our planet-hunting techniques find large planets much more easily. As our instruments and techniques improve, we expect to be able to find many more Earth-sized planets.

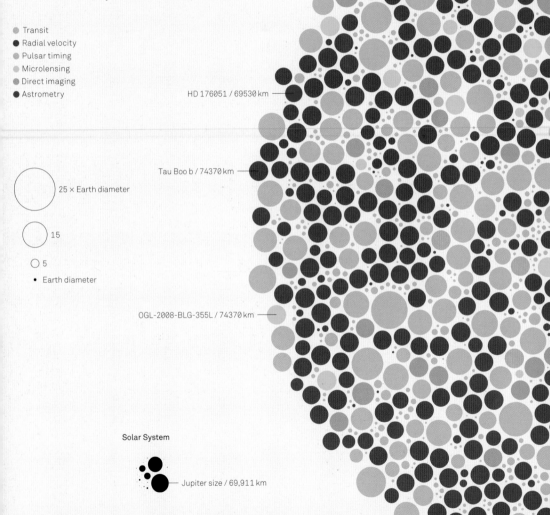

Transit
Radial velocity
Pulsar timing
Microlensing
Direct imaging
Astrometry

HD 176051 / 69530 km

25 × Earth diameter

15

5

Earth diameter

Tau Boo b / 74370 km

OGL-2008-BLG-355L / 74370 km

Solar System

Jupiter size / 69,911 km

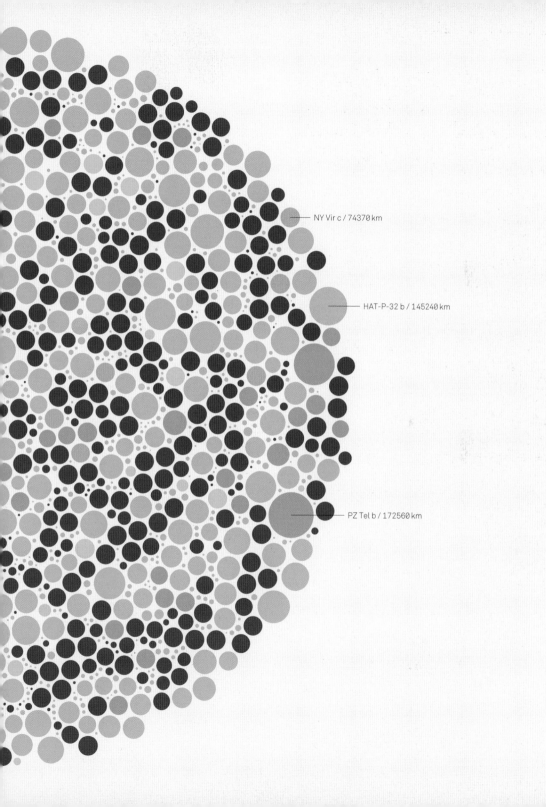

NY Vir c / 74370 km

HAT-P-32 b / 145240 km

PZ Tel b / 172560 km

Planetary systems

We now know of over 1,180 planetary systems. Many are very unlike our Solar System having super-Jupiters close in to their parent star. Being so close to their stars those planets will be intensely hot and inhospitable.

The main question we'd like to answer is 'are there any that we (or alien life) could live on?'. Our first requirement for a hospitable planet is for it to be at a suitable distance from its star that it isn't too hot or too cold and water can exist as a liquid. This livable range is known as the Habitable Zone. Any closer the water will boil off. Any further it turns to ice and makes it hard for life. In recent years we have found several solar systems that appear to have planets sitting in this zone. So far, these have tended to be around smaller, cooler stars than our Sun.

○ Habitable zone
⋯ Size of Earth's orbit

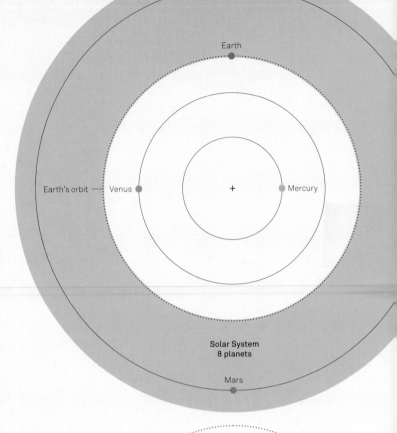

Earth

Earth's orbit — Venus   +   Mercury

Solar System
8 planets

Mars

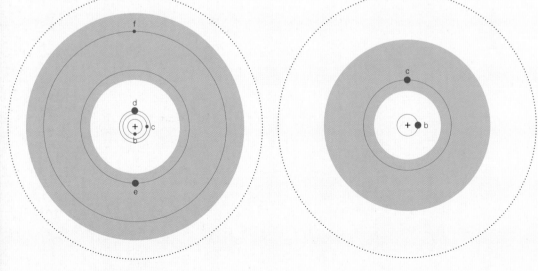

f

d
+ c
b

e

Kepler-62
5 planets

c

+ b

Kepler-283
2 planets

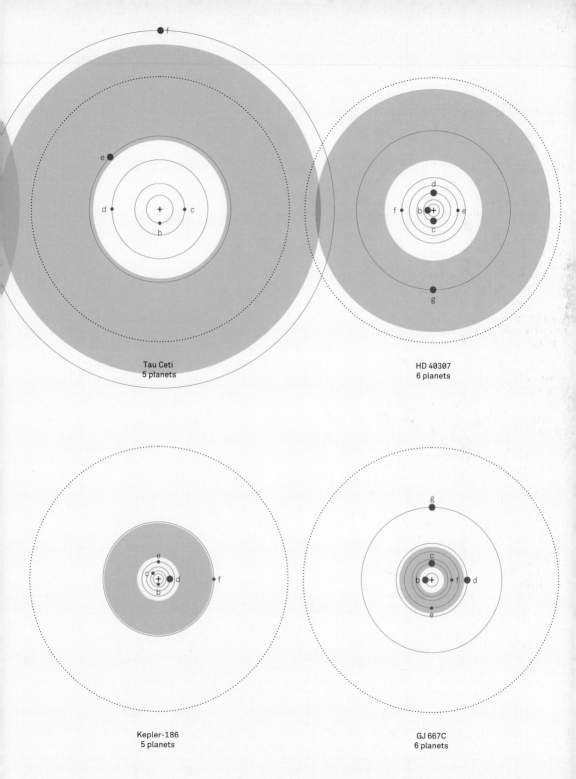

Tau Ceti
5 planets

HD 40307
6 planets

Kepler-186
5 planets

GJ 667C
6 planets

Earth-like

Planets in the *Habitable Zone* have survivable surface temperatures but that doesn't mean they are identical to the Earth. They could easily be much more massive – like Jupiter – or tiny like Ceres. For humans it will also need easily accessible water, about the right gravity and probably a solid surface. We don't have an answer to the water question at the moment but, for many exoplanets, we can calculate an approximate surface temperature, surface gravity, and if it is rocky or gassy. We can use those numbers to get a rough idea as to how Earth-like planets might be. We've certainly come a long way over the past 25 years.

● Solar System object
◖ Exoplanet

● Earth diameter
○ 3 × Earth diameter
○ 10 × Earth diameter
○ 30 × Earth diameter

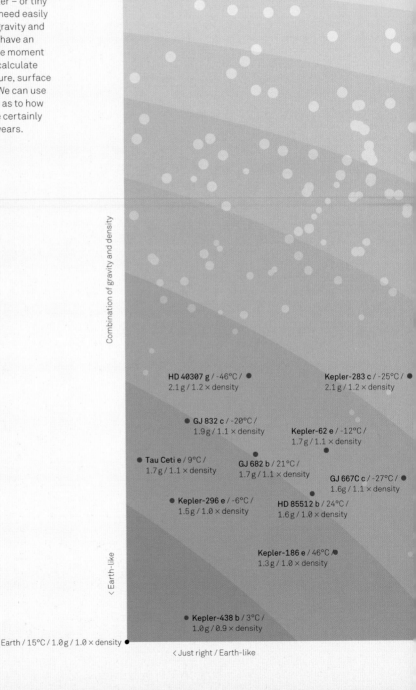

Not like Earth >

Combination of gravity and density

< Earth-like

HD 40307 g / -46°C / ●
2.1 g / 1.2 × density

Kepler-283 c / -25°C / ◖
2.1 g / 1.2 × density

GJ 832 c / -20°C /
1.9 g / 1.1 × density

Kepler-62 e / -12°C /
1.7 g / 1.1 × density

Tau Ceti e / 9°C /
1.7 g / 1.1 × density

GJ 682 b / 21°C /
1.7 g / 1.1 × density

GJ 667C c / -27°C / ◖
1.6 g / 1.1 × density

Kepler-296 e / -6°C /
1.5 g / 1.0 × density

HD 85512 b / 24°C /
1.6 g / 1.0 × density

Kepler-186 e / 46°C ◖
1.3 g / 1.0 × density

Kepler-438 b / 3°C /
1.0 g / 0.9 × density

Earth / 15°C / 1.0 g / 1.0 × density ●

< Just right / Earth-like

Ceres / -106°C / 0.0g / 0.4 × density

Hygiea •

Iapetus •

Enceladus •

Titania •

Haumea •

Triton •

Callisto •

Titan •

Ganymede •

Europa •

Io •

Moon / -53°C / 0.2g /
0.6 × density

Jupiter / -121°C / 2.6g / 0.2 × density ●

Saturn / -139°C / 1.1g / 0.1 × density ●

Mars / -46°C / 0.4g /
0.7 × density

Mercury / 167°C / 0.4g /
1.0 × density

Uranus / -197°C / ●
0.9g / 0.2 × density

Neptune / -201°C / ●
1.1g / 0.3 × density

Kepler-62 f / -72°C /
1.4g / 1.0 × density

GJ 667C f / -52°C /
1.4g / 1.0 × density

GJ 667C e / -84°C /
1.4g / 1.0 × density

Kepler-186 f / -85°C /
1.1g / 0.9 × density

Venus / 457°C / 0.9g /●
0.9 × density

Temperature

Too hot or too cold / Not like Earth >

Are we alone?

As of yet, we have not found life elsewhere in the Universe. That doesn't mean that we are alone in the cosmos; space is big and we've only just started searching.

In 1961, radio astronomer Frank Drake put forward an equation to estimate the number of intelligent civilisations that we could communicate with in the Galaxy. The equation is a way to help us think about what is needed for life to exist. It starts out by working out the chances of the right kind of star and planet for life to get started on. It then attempts to estimate the chances of life developing to the point where it decides to let others know of its existence. Over the years since the equation was proposed, we've been able to get better estimates for some of the numbers. However, the number with the largest uncertainty is the length of time a civilisation exists for.

We've only been in the communicative state for under a century. How long can we exist without destroying ourselves?

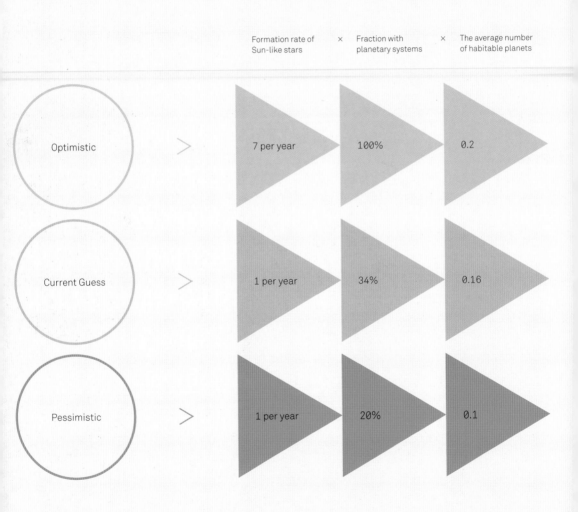

		Formation rate of Sun-like stars	×	Fraction with planetary systems	×	The average number of habitable planets
Optimistic	>	7 per year		100%		0.2
Current Guess	>	1 per year		34%		0.16
Pessimistic	>	1 per year		20%		0.1

Limitations

The equation makes a few assumptions, not least that a solar-type star and a planet are required for life. This could be too restrictive. In recent years we've found life existing around hydrothermal vents at the bottom of the oceans. These vents provide energy and nutrients to support life that may be entirely independent of the Sun. Similar processes could exist on the moons Enceladus and Europa, with energy provided from tidal interactions with their parent planets.

Does life even need a planet or moon? Experiments have shown that microscopic tardigrades are able to survive in a huge range of temperatures, high doses of radiation, and even the vacuum of space. Basic life can cope with much more extreme environments than we had previously assumed.

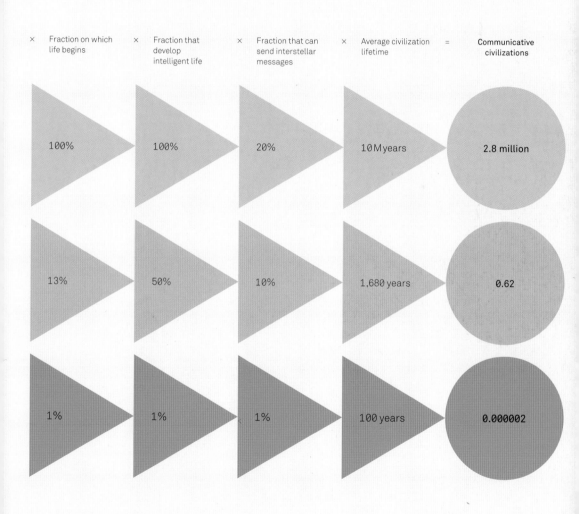

× Fraction on which life begins	× Fraction that develop intelligent life	× Fraction that can send interstellar messages	× Average civilization lifetime	= Communicative civilizations
100%	100%	20%	10 M years	2.8 million
13%	50%	10%	1,680 years	0.62
1%	1%	1%	100 years	0.000002

Postcards from Earth

When we launched spacecraft that would eventually leave the Solar System, we included messages to any extra-terrestrials that found them. The two Pioneer spacecraft *(Pioneers 10 & 11)* included gold-anodized aluminium plaques with artwork by Linda Salzman Sagan, Carl Sagan, and Frank Drake.

The two Voyager probes *(Voyagers 1 & 2)* took golden records. None of our probes will reach close to other star systems for many tens of thousands of years. Will ET be able to decipher the messages? Can you?

Pioneer plaque

Width 229 mm × Height 152 mm

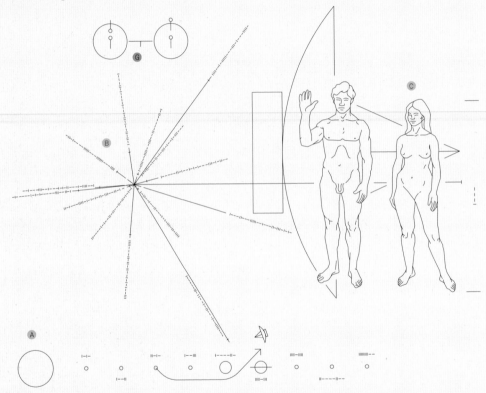

A Solar System

A representation of the Solar System with a curved line for the path of the spacecraft. The 'arrowhead' on the trajectory has been controversial as it is symbol from our hunter-gatherer past and may not be recognised. The objects depicted may be confusing given that their positions are not to scale and they aren't all the largest objects.

B Pulsar Map

This shows the positions of 14 pulsars relative to the Sun with the lengths of the lines being the relative distances. The symbols along the lines are a binary representation of each pulsar's rotation frequency relative to the hydrogen atom at the time of launch. The horizontal line towards the right shows the position of the centre of the Galaxy.

C Woman and Man

Representations of humanity and the spacecraft on the same scale. The depictions of humans may be harder to decipher for extra-terrestrials.

Voyager golden record
Diameter 305 mm (12 inch)

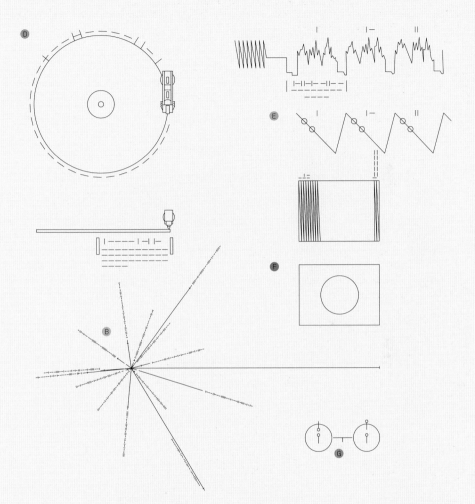

D Playing instructions

A plan and side view of the golden record with the enclosed stylus in position. Around the record is a representation, in binary, of the playing speed compared to the time associated with a hydrogen atom.

E Video signal

These illustrations show the mechanics of recreating the video signal. The first part shows the waveform generated by playing the record. The rectangles represent the way to build up an image with binary code so that there are 512 vertical lines. The bottom rectangle shows a circle which is the first image that will be shown if ET gets it to work.

F Content

The video contains 116 images as well as various sounds of Earth including music by Bach and Chuck Berry.

G Hydrogen Atom

This shows an energy transition of the hydrogen atom: the most common element in the Universe. It is associated with a frequency (1420.406 MHz) and a wavelength (21 cm) to provide scale for other parts of the image.

Earth calling

In 1974, radio astronomers at the Arecibo radio telescope transmitted an image in the direction of globular star cluster *M13*. The cluster, containing around 300,000 stars, is 25,000 light years away in the direction of Hercules.

The transmission switched between two slightly different frequencies, near 2,380 MHz, to create a binary code. In total, 1,679 binary digits were sent; a number chosen because it is the product of two prime numbers. The hope was that extra-terrestrials would recognise that property

of this number and would then arrange the message in a 23 × 73 pixel image. Even if ET work out how to display the image, they would still have to work out what it means. We just don't know how difficult that might be for them. It certainly isn't easy for humans to work out and it is *our* message.

Unfortunately, the cluster will move over the next 25,000 years so the message will miss any civilisations living there. We certainly shouldn't expect a reply anytime soon.

The numbers one to ten in a binary representation.

A binary representation of the atomic numbers of the elements making up DNA: hydrogen, carbon, nitrogen, oxygen and phospherous.

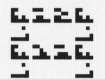

Formulas representing the chemical building blocks of DNA.

An image of DNA double helix with the number of nucleotides in the human genome encoded in binary in the middle.

A picture of a human connected to the DNA strands. The number of humans (4 Bn at the time of the message) is included on the right.

The solar system showing the location of Earth raised and centred on the human.

A picture of the Arecibo dish in Puerto Rico with the size of the antenna in binary.

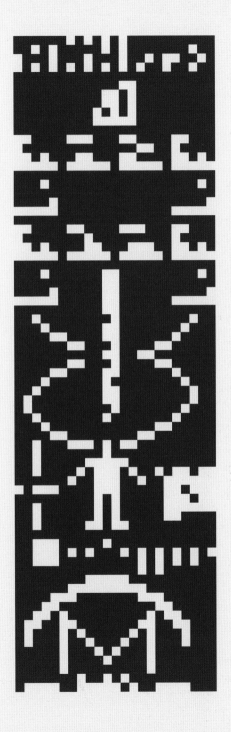

Light spheres

A fraction of the signal from all our TV and radio broadcasts leaks out into space. Unimpeded, these signals head off at the speed of light in an expanding sphere around us. In theory, alien civilisations with sensitive radio telescopes could pick up our broadcasts and listen in. Avid alien fans' knowledge of our history would be behind by a number of years determined by how far away they are.

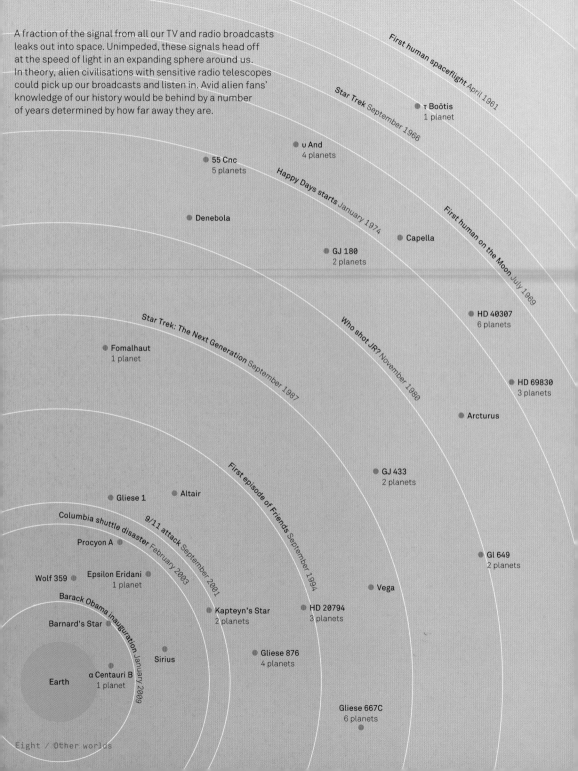

First human spaceflight April 1961

Star Trek September 1966

τ Boötis
1 planet

υ And
4 planets

55 Cnc
5 planets

Happy Days starts January 1974

Denebola

Capella

GJ 180
2 planets

First human on the Moon July 1969

HD 40307
6 planets

Who shot JR? November 1980

Star Trek: The Next Generation September 1987

Fomalhaut
1 planet

HD 69830
3 planets

Arcturus

First episode of Friends September 1994

GJ 433
2 planets

Gliese 1

Altair

Columbia shuttle disaster February 2003

9/11 attack September 2001

Procyon A

Gl 649
2 planets

Wolf 359

Epsilon Eridani
1 planet

Vega

Barack Obama inauguration January 2009

Kapteyn's Star
2 planets

HD 20794
3 planets

Barnard's Star

Sirius

Gliese 876
4 planets

Earth

α Centauri B
1 planet

Gliese 667C
6 planets

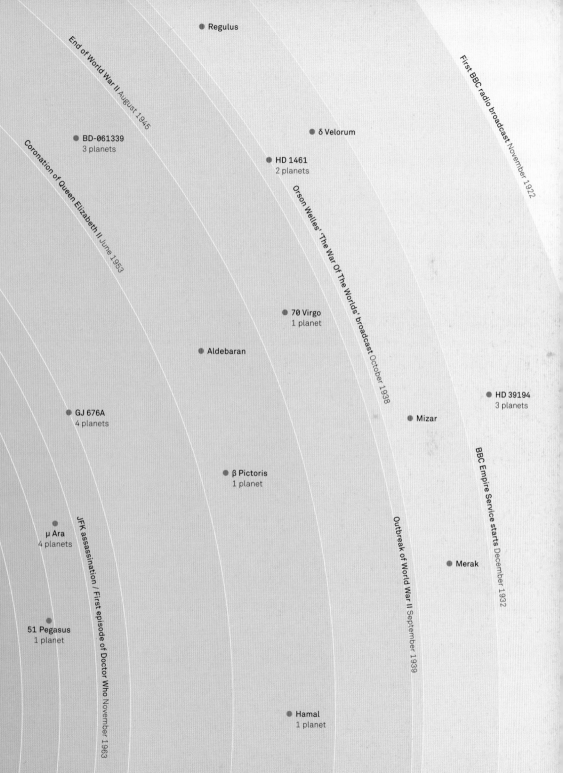

Regulus

End of World War II August 1945

First BBC radio broadcast November 1922

δ Velorum

BD-061339
3 planets

HD 1461
2 planets

Coronation of Queen Elizabeth II June 1953

Orson Welles' 'The War Of The Worlds' broadcast October 1938

70 Virgo
1 planet

Aldebaran

HD 39194
3 planets

GJ 676A
4 planets

Mizar

β Pictoris
1 planet

μ Ara
4 planets

JFK assassination / First episode of Doctor Who November 1963

Merak

BBC Empire Service starts December 1932

Outbreak of World War II September 1939

51 Pegasus
1 planet

Hamal
1 planet

Messaging extra-terrestrial intelligence

Whereas the search for extra-terrestrial intelligence (SETI) is the passive activity of listening for messages from alien life, messaging extra-terrestrial intelligence (METI) is the opposite; telling the Universe at large 'We are here! You are not alone!'. Aside from the weak and incidental transmissions of our TV and radio broadcasts, over the years there have been several deliberate attempts to send messages to specific targets. These have come from academic and commercial sources. Our first contact messages have certainly varied hugely in content.

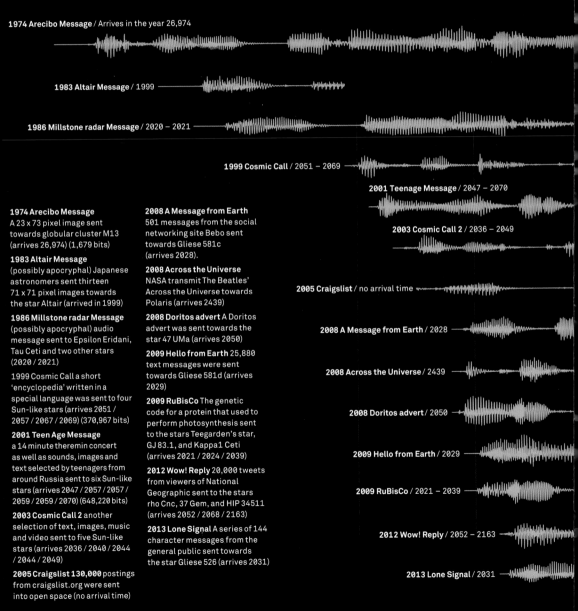

1974 Arecibo Message / Arrives in the year 26,974

1983 Altair Message / 1999

1986 Millstone radar Message / 2020 – 2021

1999 Cosmic Call / 2051 – 2069

2001 Teenage Message / 2047 – 2070

2003 Cosmic Call 2 / 2036 – 2049

2005 Craigslist / no arrival time

2008 A Message from Earth / 2028

2008 Across the Universe / 2439

2008 Doritos advert / 2050

2009 Hello from Earth / 2029

2009 RuBisCo / 2021 – 2039

2012 Wow! Reply / 2052 – 2163

2013 Lone Signal / 2031

1974 Arecibo Message
A 23 x 73 pixel image sent towards globular cluster M13 (arrives 26,974) (1,679 bits)

1983 Altair Message
(possibly apocryphal) Japanese astronomers sent thirteen 71 x 71 pixel images towards the star Altair (arrived in 1999)

1986 Millstone radar Message
(possibly apocryphal) audio message sent to Epsilon Eridani, Tau Ceti and two other stars (2020 / 2021)

1999 Cosmic Call a short 'encyclopedia' written in a special language was sent to four Sun-like stars (arrives 2051 / 2057 / 2067 / 2069) (370,967 bits)

2001 Teen Age Message
a 14 minute theremin concert as well as sounds, images and text selected by teenagers from around Russia sent to six Sun-like stars (arrives 2047 / 2057 / 2057 / 2059 / 2059 / 2070) (648,220 bits)

2003 Cosmic Call 2 another selection of text, images, music and video sent to five Sun-like stars (arrives 2036 / 2040 / 2044 / 2044 / 2049)

2005 Craigslist 130,000 postings from craigslist.org were sent into open space (no arrival time)

2008 A Message from Earth
501 messages from the social networking site Bebo sent towards Gliese 581c (arrives 2028).

2008 Across the Universe
NASA transmit The Beatles' Across the Universe towards Polaris (arrives 2439)

2008 Doritos advert A Doritos advert was sent towards the star 47 UMa (arrives 2050)

2009 Hello from Earth 25,880 text messages were sent towards Gliese 581d (arrives 2029)

2009 RuBisCo The genetic code for a protein that used to perform photosynthesis sent to the stars Teegarden's star, GJ 83.1, and Kappa1 Ceti (arrives 2021 / 2024 / 2039)

2012 Wow! Reply 20,000 tweets from viewers of National Geographic sent to the stars rho Cnc, 37 Gem, and HIP 34511 (arrives 2052 / 2068 / 2163)

2013 Lone Signal A series of 144 character messages from the general public sent towards the star Gliese 526 (arrives 2031)

To send or not to send?

That is a question with no global consensus. Some, including the likes of Steven Hawking, worry that this could encourage advanced, hostile, aliens to visit us with ill intent. Perhaps these civilisations would find us anyway so this concern may be moot. Others have called for a moratorium on further messages until we've discussed the implications globally. Alternatively, making first contact with alien life would be an historic and inspiring event that would answer the fundamental question of 'Are we alone?'.

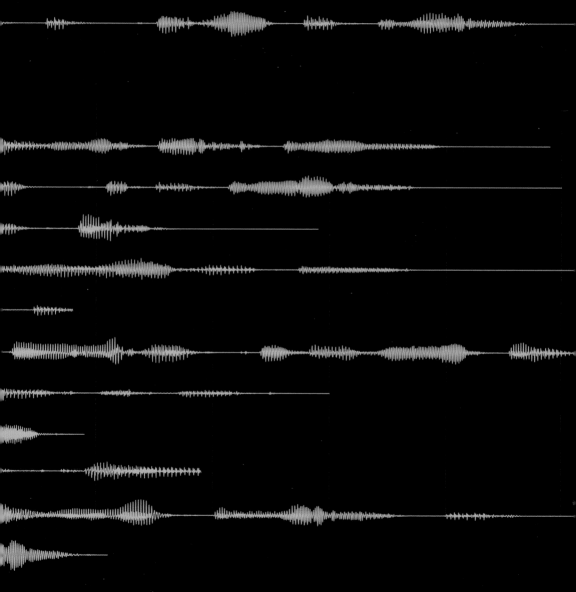

Nine / Miscellaneous

Relativistic effects

In 1905 Albert Einstein published his Special Theory of Relativity, which showed that measurements of time and distance change for observers moving at speed.

He followed it up in 1915 with his General Theory of Relativity, which showed the effect of gravity on light. These effects are not normally noticeable on human scales, but there are situations where they become important.

Time dilation

How old are you, exactly? Einstein showed that time is not a constant. Time runs more slowly for people who are moving very fast, or who are in a gravitational field, though the differences are very small. In 1971 Joseph Hafele and Richard Keating took atomic clocks on commercial round-the-world plane journeys in opposite directions.

Younger due to speed

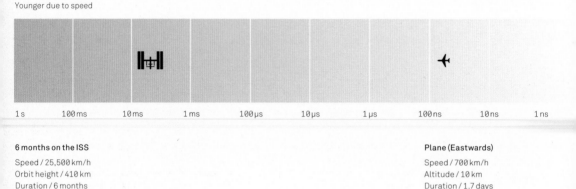

| 1 s | 100 ms | 10 ms | 1 ms | 100 µs | 10 µs | 1 µs | 100 ns | 10 ns | 1 ns |

6 months on the ISS
Speed / 25,500 km/h
Orbit height / 410 km
Duration / 6 months

Plane (Eastwards)
Speed / 700 km/h
Altitude / 10 km
Duration / 1.7 days

Black holes

The General Theory of Relativity predicts that black holes should exist. While we have never seen any directly, there is very good evidence for them. The motions of stars near the Galactic Centre must be around an object four million times the mass of the Sun, but which is completely invisible.

Gravitational lensing

Einstein's gravitation theories also predicted that massive objects distort the fabric of space, and that light passing near such objects would be bent. That can either make background objects appear to be in different places or, if the object is massive enough, create multiple images.

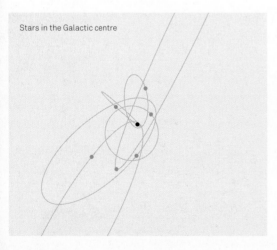

Stars in the Galactic centre

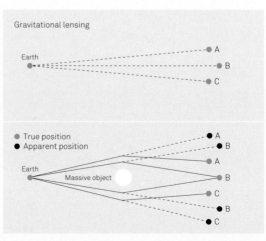

Gravitational lensing

They showed that the effect is real, with the difference being because the Earth's surface is moving at 1,600 kilometres per hour at the equator. Satellite navigation systems rely on very precise timings, and the tiny difference would result in positions being wrong by over 100 metres per day if not accounted for.

Twin paradox

The differences are more noticeable when considering future interplanetary travel. At one tenth of the speed of light an explorer could travel to Neptune and back in less than four days, and would return 25 minutes younger than if they had stayed on Earth.

Older due to gravity

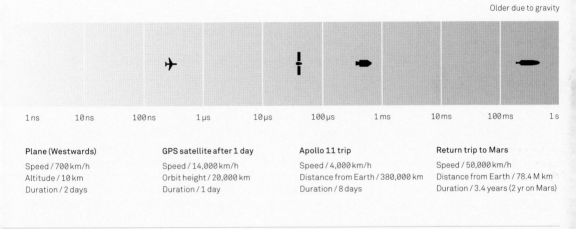

| 1 ns | 10 ns | 100 ns | 1 µs | 10 µs | 100 µs | 1 ms | 10 ms | 100 ms | 1 s |

Plane (Westwards)

Speed / 700 km/h
Altitude / 10 km
Duration / 2 days

GPS satellite after 1 day

Speed / 14,000 km/h
Orbit height / 20,000 km
Duration / 1 day

Apollo 11 trip

Speed / 4,000 km/h
Distance from Earth / 380,000 km
Duration / 8 days

Return trip to Mars

Speed / 50,000 km/h
Distance from Earth / 78.4 M km
Duration / 3.4 years (2 yr on Mars)

In 1919, Arthur Eddington observed stars during a solar eclipse and found them to be in slightly different locations than normal. Although the apparent bending of the light was a tiny effect, this was the first observational proof of Einstein's theory.

Massive clusters of galaxies act like gravitational lenses, magnifying and distorting the light from more distant objects. The background objects appear as arcs in the images, allowing us to study some of the most distant galaxies in the Universe.

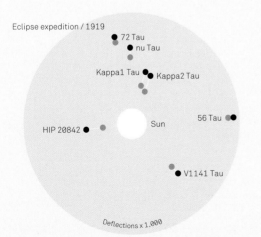

Eclipse expedition / 1919

72 Tau
nu Tau
Kappa1 Tau ● ● Kappa2 Tau
56 Tau
HIP 20842 ● Sun
V1141 Tau

Deflections x 1,000

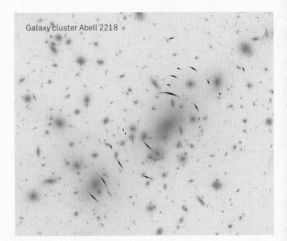

Galaxy cluster Abell 2218

Pictures of the day

The Astronomy Picture of the Day (APOD) website has been online since 16 June 1995. Every day it shows a different image of space along with a brief explanation. We've broken down the type of objects that have been included in APOD over its 20 year history.

These days APOD also appears on social media. Recent analysis has shown that the most popular types of APOD images shared by Google Plus users are of planets and moons. Skyscapes are the most popular amongst Facebook and Twitter users.

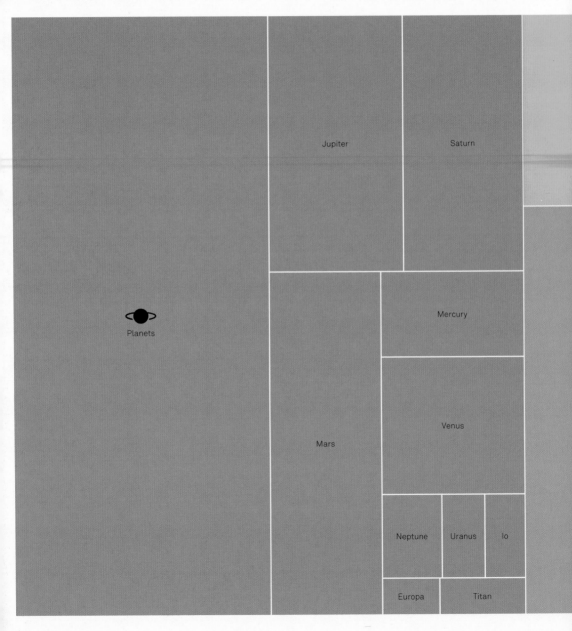

Planets

Jupiter

Saturn

Mercury

Venus

Mars

Neptune

Uranus

Io

Europa

Titan

Stars and clusters

Eagle Nebula

Melotte 1

Interplanetary bodies
and dwarf planets

Pluto

Comet
Hale-Bopp

Comet
Hyakutake

Lagoon Nebula

Trifid Nebula

Nebulae

Orion Nebula

Galaxies

Andromeda Galaxy

Cigar Galaxy

Bode's Galaxy

Triangulum
Galaxy

Sky phenomenon

Orion

Solar Eclipse

Cosmology

Technology

Water worlds

The Earth is the only body in the Solar System with liquid water on its surface, but it's not the only place where we find water. In most cases, such as on Mars, this water is in the form of ice, either on the surface or within the rock layers just beneath the surface. The sub-surface ice does melt for short periods when heated by sunlight, causing rivulets to run down crater walls before evaporating in the low atmospheric pressure.

But there is water elsewhere in the Solar System, in the most unlikely of places. Jupiter's moon Europa has a thick layer of ice covering its surface, as do many moons in the outer Solar System. The surface is constantly being replaced, and we know that beneath it lies an ocean of water. This ocean is prevented from freezing by internal heat generated by tides caused by Jupiter. Europa may contain more water than all the Earth's oceans put together.

Saturn's moon Enceladus also has liquid water beneath its surface. This water has been observed escaping from the moon's south pole in salty geysers. We now think that that many of the larger moons in the outer Solar System have sub-surface oceans, though their depths – and volumes – are far from certain.

Earth / diameter 12,742 km

Europa / diameter 3,122 km

3 billion km³ Ice crust
with sub-surface ocean

1.4 billion km³
Ice, water and vapour

7.5 million km³ Ice with
water vapour geysers

Enceladus / diameter 504 km

5,000 km³ Ice at poles and
subsurface permafrost

Mars / diameter 6,779 km

Density

Space is very empty. The density of the gas between the stars is a thousand million millionth (quadrillionth) the density of air. But the objects in space can be very dense indeed. The centre of the Sun is 50 times denser than rock, and neutron stars are almost unimaginably dense. Since densities are hard to imagine, we compare how much standard volumes of a substance would weigh. The standard volume used here is a bucket.

↓ has the same mass as a bucket of →

	Interstellar cloud	Air	Water	Rock

Virus 10 quadrillionths of a gram

Smallest bacterium 300 quadrillionths of a gram

Fly 2.5 thousanths of a gram

Teaspoon of water 5.0 g

Laptop 2 kg
Water 10 kg

Lion 130 kg

Car 1 tonne

HGV 7.5 tonnes

Semi-trailer truck 44 tonnes

747 plane 333 tonnes

Space Shuttle at launch 2,041 tonnes

Titanic 41,000 tonnes

Supertanker (ship) 420,000 tonnes

Mount Everest 161 billion tonnes

Antarctic ice sheet 30 quadrillion tonnes

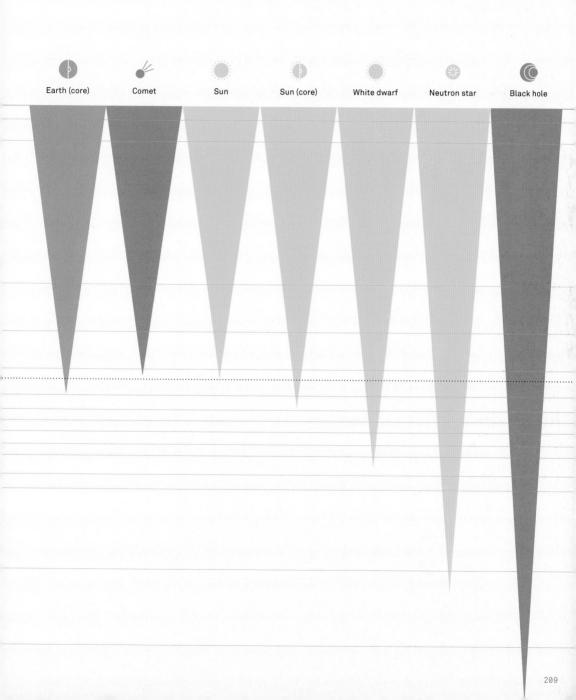

Earth (core)   Comet   Sun   Sun (core)   White dwarf   Neutron star   Black hole

Building blocks

The Universe is composed almost completely of hydrogen and helium, formed within the first few minutes after the Big Bang. Successive generations of stars have created heavier elements, though only in relatively small amounts, and these are present in the Sun and Solar System.

As the planets were forming, the lighter elements were pushed outwards, leaving elements such as oxygen, carbon and silicon to form the inner planets. Heavy elements such as iron have sunk into the Earth's core, leaving a crust made primarily of silicon and oxygen.

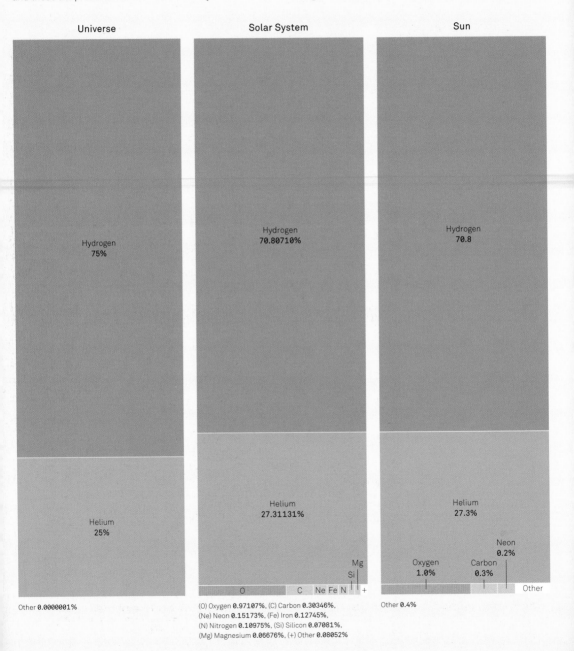

Universe

Hydrogen
75%

Helium
25%

Other 0.0000001%

Solar System

Hydrogen
70.80710%

Helium
27.31131%

Mg
Si
O   C   Ne Fe N +

(O) Oxygen 0.97107%, (C) Carbon 0.30346%,
(Ne) Neon 0.15173%, (Fe) Iron 0.12745%,
(N) Nitrogen 0.10975%, (Si) Silicon 0.07081%,
(Mg) Magnesium 0.06676%, (+) Other 0.08052%

Sun

Hydrogen
70.8

Helium
27.3%

Neon
0.2%
Oxygen
1.0%
Carbon
0.3%
Other

Other 0.4%

The oceans are thought to have been created by impacts from asteroids and comets, returning some lighter elements to the Earth's surface. We are made of the same elements as our planet, just in slightly different proportions. Our DNA is made of carbon, oxygen, nitrogen, hydrogen and phosphorus, while calcium is the key to strong bones. Apart from hydrogen, all these elements were formed in stars – we really are stardust!

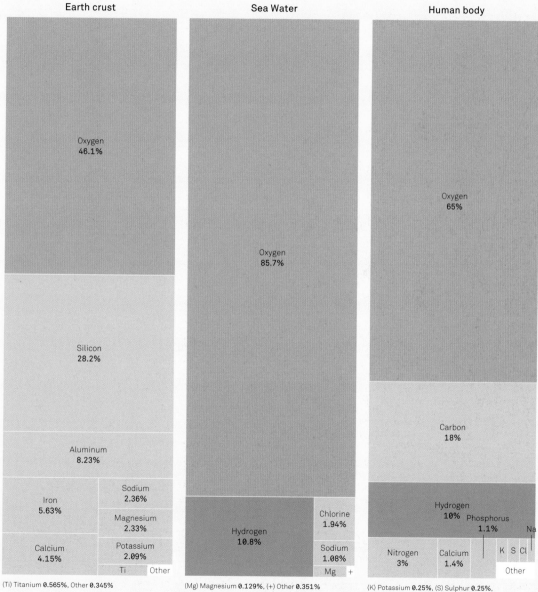

Earth crust

Oxygen
46.1%

Silicon
28.2%

Aluminum
8.23%

Iron
5.63%

Sodium
2.36%

Magnesium
2.33%

Calcium
4.15%

Potassium
2.09%

Ti   Other

(Ti) Titanium 0.565%, Other 0.345%

Sea Water

Oxygen
85.7%

Hydrogen
10.8%

Chlorine
1.94%

Sodium
1.08%

Mg   +

(Mg) Magnesium 0.129%, (+) Other 0.351%

Human body

Oxygen
65%

Carbon
18%

Hydrogen
10%

Phosphorus
1.1%

Na

Nitrogen
3%

Calcium
1.4%

K S Cl

Other

(K) Potassium 0.25%, (S) Sulphur 0.25%,
(Cl) Chlorine 0.15%, (Na) Sodium 0.15%,
Other 0.7%

How long is a day?

Did it feel like a long day today? It probably was. A standard day is defined as the time for one rotation of the planet and is taken as 86,400 seconds. However, the Earth's spin is not constant; sometimes it can be slightly faster and at other times slower. For example, the 2004 Indonesian earthquake brought a large section of tectonic plate slightly inwards and shortened the day by 2.7 microseconds. Conversely, the tides caused by the Moon are slightly slowing the Earth and increase the day by about 15 – 20 microseconds per year. The Earth Orientation Centre of the International Earth Rotation and Reference Systems Service (IERS) uses atomic clocks and radio observations of distant quasars to measure the precise length of the day allowing us to see how

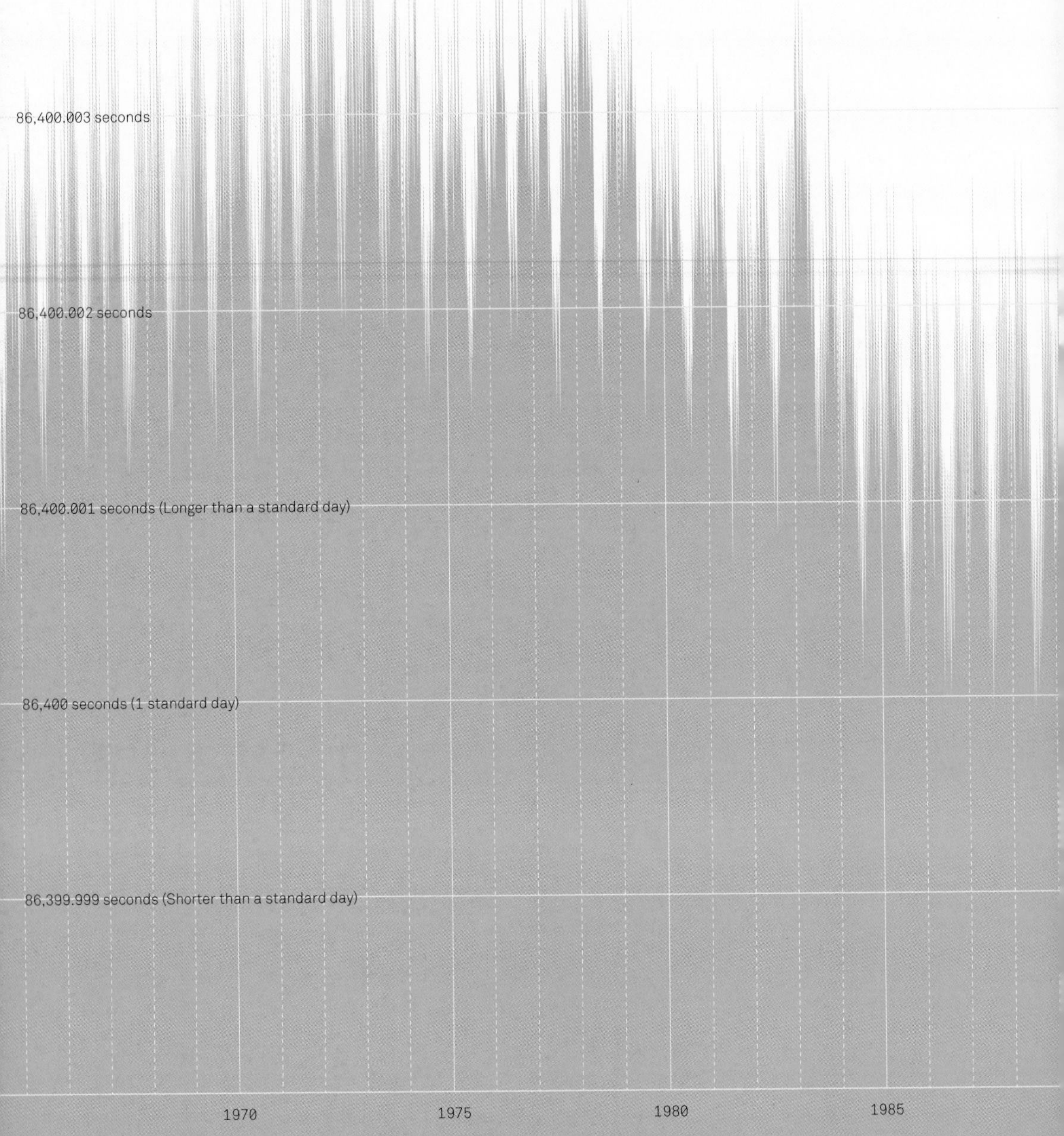

86,400.003 seconds

86,400.002 seconds

86,400.001 seconds (Longer than a standard day)

86,400 seconds (1 standard day)

86,399.999 seconds (Shorter than a standard day)

1970   1975   1980   1985

it changes over time. Changes over decades are thought to be due to processes within the Earth's core. Changes on timescales less than two years are largely down to our atmosphere affecting the rotation of the planet.

The longest day on record was on the 12 April 1972 and lasted 4.36 milliseconds longer than the standard day. With the day being a millisecond or so longer than our definition, this slowly adds up over time.

After a few hundred days, the day will be wrong by as much as a second. To keep atomic time matched to the Earth, we occasionally add leap seconds. The IERS have added 25 leap seconds since 1972.

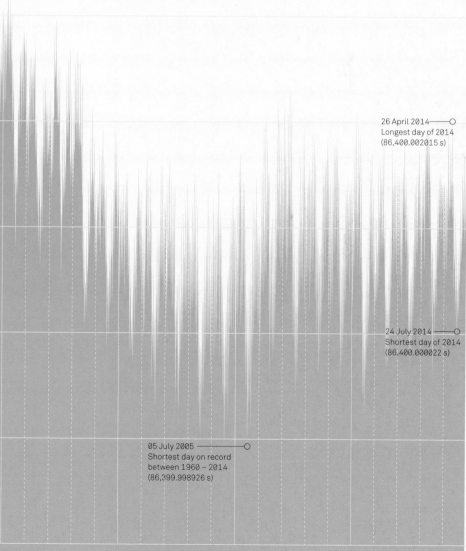

26 April 2014 ———O
Longest day of 2014
(86,400.002015 s)

24 July 2014 ———O
Shortest day of 2014
(86,400.000022 s)

05 July 2005 ———O
Shortest day on record
between 1960 – 2014
(86,399.998926 s)

1995   2000   2005   2010 2011 2012 2013 2014

Identified flying objects

We've all seen things in the sky that seemed strange at first glance. Simple things such as Venus, the International Space Station, and Chinese lanterns can all seem quite mysterious if you aren't used to them. People sometimes contact local observatories, university astronomy departments, or even the police to find out what they are looking at. Usually, a few questions, a bit of deduction and a process of elimination will provide a likely explanation for these unknown aerial objects. The truth is out there. It just might be an uplit seagul.

Messier catalogue

Charles Messier was a French astronomer who moved
to Paris when he was 14 years old. He wanted to find new
comets but would often re-discover other fuzzy objects
in the sky such as nebulae and star clusters.

To stop himself wasting time on these objects he created
a catalogue of their positions. His catalogue – a simple list
of things he didn't want to observe – has become one of his
lasting legacies.

* Group of stars
◇ Open cluster
◆ Globular cluster
◎ Nebula
⊗ Supernova remnant
● Galaxy
ᖚ Spiral galaxy
↻ Barred galaxy
◣ Elliptical galaxy
⚭ Interacting galaxy

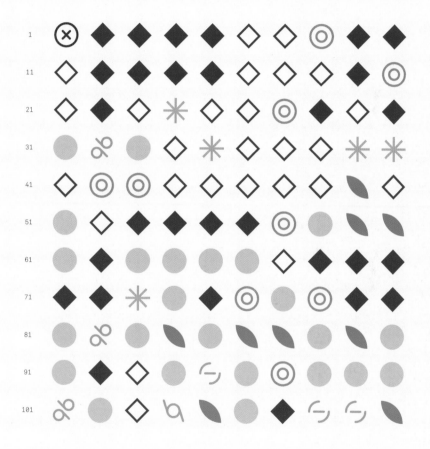

New General Catalogue

In the late 19th Century, the Royal Astronomical Society asked John Dreyer to compile a New General Catalogue (NGC) of nebulae and clusters. This was a large catalogue curated from previous catalogues and observations of astronomers from many different countries using many different telescopes.

The catalogue was arranged by angle on the sky so nearby catalogue entries can usually be observed at a similar time of night. In the regions *NGC 1700 – NGC 2500* and *NGC 6300 – NGC 7100* the catalogue cuts along the plane of the Galaxy so these parts show far more star clusters and nebulae than others. Since publication, nearly 60 of the objects have been found not to exist; they have either been lost or were mistakes.

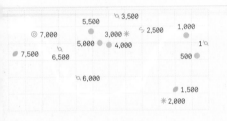

✕ Star
✳ Group of stars
◇ Open cluster
◆ Globular cluster
◎ Nebula
⊗ Supernova remnant
⬤ Galaxy
ʖ Spiral galaxy
ʂ Barred galaxy
◗ Lenticular galaxy
◣ Elliptical galaxy
⅋ Interacting galaxy
　 Lost / Doesn't exist (gap)

Unsung heroes of astronomy

Few astronomers become well known for their discoveries.
Here are a selection who have increased our knowledge of
the Universe but remain relatively unknown.

Aristarchus of Samos
310 – 230 BCE

Greek astronomer who first proposed
the Sun as the centre of the Solar System.
He also suggested that the stars were
other suns. He attempted to measure the
relative distances of the Moon and Sun.

Thomas Harriot
1560 – 1621

English astronomer who was the first
person to observe the Moon through
a telescope.

Maria Margarethe Kirch
1670 – 1720

German astronomer who studied the Aurora
Borealis (northern lights), planetary
conjunctions and created calendars. She was
the first woman to discover a comet although
the credit was taken by her husband.

Caroline Herschel
1750 – 1848

She was the first woman credited with
the discovery of a comet in her own right.
Discovered eight comets. In 1787 she was
paid a salary by King George III to work
as an astronomical assistant.

John Goodricke
1764 – 1786

Dutch-English amateur astronomer
proposed the concept of an eclipsing
binary star to explain his observations
of the star Algol.

Urbain Le Verrier
1811 – 1877

He studied perturbations in the orbit of
Uranus. Using mathematics he determined
that these were due to an unknown body
and sent his predictions of its location to
Johann Galle at the Berlin Observatory.
Galle found the planet Neptune within an
hour of starting to look.

Angelo Secchi
1818 – 1878

Italian astronomer who invented the
heliospectrograph for studying the
spectrum of light from the Sun. He proved
that prominences observed during eclipses
were part of the Sun. He also discovered
three comets and was the first to describe
'channels' (canali) on Mars.

Williamina Fleming
1857 – 1911

Scottish-American astronomer who
discovered about 40% of the novae
that were then known.

Annie Jump Cannon
1863 – 1941

American astronomer who created a
classification scheme for stars that split
them into the classes O, B, A, F, G, K, and M.

Annie Scott Dill Maunder
1868 – 1947

Irish astronomer who worked at the Royal Observatory Greenwich observing the Sun. She was an expert in eclipse photography. Together with her husband she discovered the minimum in sunspot numbers now known as the 'Maunder Minimum'.

Henrietta Swan Leavitt
1868 – 1921

American astronomer who discovered Cepheid variable stars which provided a standard candle to measure the Universe.

Georges Lemaître
1894 – 1966

Belgian cosmologist and priest who proposed the expansion of the Universe, first estimated the Hubble constant, and proposed that the Universe began with an explosion at a point in time.

Fritz Zwicky
1898 – 1974

Swiss astronomer who made contributions to many areas of astronomy. He found 123 supernovae and even helped coin the word for them. He predicted the existence of gravitational lenses due to galaxy clusters 42 years before the first one was found. He was also the first person to observe the effects of dark matter.

Celia Payne-Gaposchkin
1900 – 1979

As a 25 year old student, Payne's doctoral dissertation argued that the majority of the Sun, stars and the Universe was composed of hydrogen. Although initially dismissed, she was later shown to be correct.

Ruby Payne-Scott
1912 – 1981

Australian astronomer who was the first female radio astronomer. She extensively studied the Sun and discovered various types of radio burst. She played a large part in the first-ever radio interferometer.

Grote Reber
1911 – 2002

American who built the first parabolic dish radio telescope. He conducted the first radio sky map identifying the galaxy and also discovering objects such as Cassiopeia A and Cynus A.

Nancy Grace Roman
1925 –

American astronomer who became NASA's first Chief of Astronomy. She oversaw the launch of three solar observatories and three astronomical satellites. She was involved in the early planning and design for the Hubble Space Telescope.

Beatrice Tinsley
1941 – 1981

A New Zealand astronomer who studied stars and galaxies. She completed her PhD at the University of Texas in only two years and her thesis was the basis for much of the subsequent study of the evolution of galaxies. During her career she worked on a huge range of topics.

Sources, notes and acknowledgements

8 / Design a Space Telescope,
Cardiff University, Wales

10 / Beischer, DE; Fregly, AR (1962) US
Naval School of Aviation Medicine /
Encyclopedia Astronautica / Dr.Kenichi IJIRI
/ Witt, P.N., et al. 1977 J. Arachnol. 4: 115-
124 / National Space Science Data Center,
NASA / Spangenberg et al. Adv Space Res.
1994;14(8):317-25 / Szewczyk, N.J. et al,
Astrobiology, Volume 5, Issue 6, pp.
690-705 / ESA.

12 / NASA Information Summaries Astronaut
Fact Book / www.spacefacts.de / NASA
History Office.

14 / NASA Information Summaries
Astronaut Fact Book / spacefacts.de /
NASA History Office.

16 / Bioastronautics Data Book: Second
Edition. NASA SP-3006 / From Quarks to
Quasars.

18-20 / Jonathan's Space Report
planet4589.org.

22 / Catalogue of Space Debris, U.S. Space
Surveillance Network (October 2014)
orbitaldebris.jsc.nasa.gov.

24 / NASA Reference Guide to the
International Space Station / China Manned
Space Engineering.

26 / Apollo 11 Press Kit (69-83K).

28 / Catalogue of Manmade Material on the
Moon, NASA History Program Office, 7-05-
12. / Apollo 11 Traverses map prepared by
the U.S. Geological Survey and published
by the Defense Mapping Agency for NASA.
/ NASA's Lunar Reconnaissance Orbiter.

30 / Wikipedia / moon.luxspace.lu.

32 / The Expensive Hardware Lob,
David Gore.

34 / JPL Horizons, Giorgini, J.D. et al, 'JPL's
On-Line Solar System Data Service', Bulletin
of the American Astronomical Society 28(3),
1158 (1996).

36 / NASA HQ / zarya.info / Unmanned
Spaceflight.com / NASA Mars Exploration
Rovers / curiositylog.com.

40 / David A. Weintraub,
Is Pluto A Planet? (2007).

44 / Solar System Exploration, NASA.

46 / JPL Solar System Dynamics / Solar
System Exploration, NASA / Moons of
Jupiter/Moons of Saturn/Moons of Uranus/
Moons of Neptune, Wikipedia / David A.
Weintraub, Is Pluto A Planet? (2007).

48 / Eclipse Predictions by Fred Espenak
(NASA's GSFC) / Felix Verbelen.

50 / IAU Working Group for Planetary System
Nomenclature. 'Gazetteer of Planetary
Nomenclature.' / Nature 453, 1212-1215 (26
June 2008) / McGill, J. Geophys. Res., 94(B3),
2753–2759 (1989) / Oshigami & Namiki,
Icarus, Volume 190, Issue 1, p. 1-14, Sep
2007 / NASA Dawn.

52 / D. Smith et al. (2012) Science, 336,
214 / A. Aitta (2012) Icarus, 218, 967 / A.
Dziewonski & D. Anderson (1981), Physics
of the Earth and Planetary Interiors, 25, 297
/ A. Rivoldini et al. (2011) Icarus, 213, 451 /
T. Guillot et al. (1997), Icarus, 130, 534 / W.
Hubbard et al. (1991), Science, 253, 648 /
R. Weber et al. (2011), Science, 331, 309 / J.
Anderson et al. (2012), J. Geophys. Res., 106,
32963 / O. Kuskov & V. Konrod (2005), Icarus,
177, 550 / S. Vance et al. (2014), Planetary
and Space Science, 96, 62 / G. Tobie et al.
(2005), Icarus, 175, 496.

54 / COSPAR International Reference
Atmosphere / T. Cavalié et al. (2008) A&A
489, 795 & 'An introduction to Planetary
Atmospheres' (Agustin Sanchez-Lavega) /

Lellouch et al. (1988) Icarus 79, 328 / Cassini/
CIRS (L. N. Fletcher et al. (2009) Icarus 202,
543 / Orton, G. et al. (2014) Icarus 243, 494 /
L.N. Fletcher et al. (2010) A&A 514, A17.

56 / Solar System Exploration, NASA.

58 / 1999 European Asteroidal Occultation
Results / Baer & Chesley (2008) / Belton et
al (1996) / Braga-Ribas et al (2014) / Carry
(2012) / Conrad (2007) / Descamps et al
(2008) / IRAS / JPL Horizons, Giorgini, J.D. et
al, 'JPL's On-Line Solar System Data Service',
Bulletin of the American Astronomical
Society 28(3), 1158 (1996) / Kaasalainen et
al Icarus 159 369–395 (2002) / Marchis et
al (2005) / Merline et al (2013) / Millis et al
(1984) / Müller & Blommaert (2004) / RASNZ
Occultation Section / Russell et al (2012) /
Schmidt et al (2008) / Shepard et al (2008)
/ Sierks et al (2011) / Storrs et al (1999) /
Storrs et al (2005) / Tedesco et al (2002) /
Thomas et al (1994) / Thomas et al (1996) /
Thomas et al (2005) / Torppa et al (2003).

60 / IAU Minor Planet Center
minorplanetcenter.net/.

62 / JPL Small-Body Database Browser.

64 / IAU Minor Planet Center
minorplanetcenter.net / EARN NEA
Database, maintained at the Institute of
Planetary Research of the DLR, Berlin,
Germany by G. J. Hahn.

66 / Catalogue of Meteorites, Natural
History Museum, London.

68 / NASA / JPL-Caltech / UMD / NASA
Stardust / Planetary Society / ESA Giotto /
SETI Institute / NASA Comet Quest.

70 / IAU Minor Planet Center
minorplanetcenter.net/.

72 / David Levy's Guide to the Night Sky,
David H. Levy / 'The Comets of Caroline
Herschel (1750-1848), Sleuth of the Skies

at Slough', Olson & Pasachoff, Culture and Cosmos, Vol. 16, nos. 1 and 2, 2012 / Biographical Encyclopedia of Astronomers.

74 / Minor Planet Physical Properties Catalogue / IAU Minor Planet Center minorplanetcenter.net/.

78 / Kominami & Ida (2002) / Levison et al (2008) / Zwart (2009) arXiv:0903.0237 / Solar System Exploration, NASA / Cox & Loeb (2008) / Sackmann, Boothroyd & Kraemer (1993).

80 / Using ST:TOS formula of speed/c = warp^3.

84 / Observatory web sites.

86 / Reimer et al, A&A 424, 773–778 (2004) / Lord, S. D., 1992, NASA Technical Memorandum 103957 / Gemini Observatory / UKIRT/JAC / SMA/Harvard-CfA.

88-98 / Observatory web sites.

104 / BASS2000, Paris Observatory, Delbouille, Neven and Roland, 1972.

106 / SILSO data, Royal Observatory of Belgium, Brussels.

108 / ROG/USAF/NOAA Sunspot Data solarscience.msfc.nasa.gov/.

110 / The 'X-ray Flare' dataset was prepared by and made available through the NOAA National Geophysical Data Center (NGDC).

112 / JPL Horizons, Giorgini, J.D. et al, 'JPL's On-Line Solar System Data Service', Bulletin of the American Astronomical Society 28(3), 1158 (1996).

116-188 / VirtualSky, LCOGT lcogt.net/virtualsky.

120 / VirtualSky, LCOGT lcogt.net/virtualsky / Perryman et al, The Hipparcos Catalogue, A&A, 323, L49-52 (1997) / van Leeuwen, A&A, 474, 2, pp.653-664 (2007) / Harper, Brown & Guinan, AJ, 135, 4, pp 1430-1440 (2008).

122 / Harrington & Dahn, AJ, Vol. 85, p. 454-465 (1980) / Matthews, QJRAS, Vol. 35, p. 1-9 (1994) / Nidever et al, ApJS, Vol. 141, Issue 2, pp. 503-522 / Salim & Gould, ApJ, Vol. 582, Issue 2, pp. 1011-1031 / Lépine & Shara, AJ, Vol. 129, Issue 3, pp. 1483-1522 / Gontcharov, Astronomy Letters, Vol. 32, Issue 11, p.759-771 / van Leeuwen, A&A, Vol. 474, Issue 2, November I 2007, pp.653-664 / Gatewood, AJ, Vol. 136, Issue 1, p. 452 (2008) / Jenkins et al, ApJ, Vol. 704, Issue 2, pp. 975-988 (2009) / Koen et al, MNRAS, Vol. 403, Issue 4, pp. 1949-1968 (2010) / Lurie et al, AJ, Vol. 148, Issue 5, article id. 91, 12 pp. (2014).

124 / VirtualSky, LCOGT lcogt.net/virtualsky / van Leeuwen, A&A, 474, 2, pp.653-664 (2007) / Zacharias et al, VizieR On-line Data Catalog: I/322A (2012) / Roeser & Bastian, A&A Supp. 74, 3, pp. 449-451 (1988) / Perryman et al, The Hipparcos Catalogue, A&A, 323, L49-52 (1997) / Høg et al, A&A, 355, pp. L27-L30 (2000).

126 / Perryman et al, The Hipparcos Catalogue, A&A, 323, L49-52 (1997).

128 / Nordgren et al, AJ, 118, 6, pp. 3032-3038 (1999) / Ramírez & Allende Prieto, ApJ, 743, 2, pp 14 (2011) / Richichi & Roccatagliata, A&A, 433, 1, pp. 305-312 (2005) / David Darling Encyclopedia of Science / Moravveji et al, ApJ, 747, 2, pp. 7 (2012) / spacemath.gsfc.nasa.gov / Schiller & Przybilla, A&A, 479, 3, pp. 849-858 (2008) / Najarro et al, ApJ, 691, 2, pp. 1816-1827 (2009) / Perrin et al, A&A, 418, pp. 675-685 (2004) / Smith, Hinkle & Ryde, AJ, 137, 3, pp. 3558-3573 (2009) / Arroyo-Torres et al, A&A, 554, p 10 (2013).

130 / Doyle & Butler, A&A, 235, 1-2, pp. 335-339 (1990) / Demory et al, A&A, 505, 1, pp. 205-215 (2009) / Kervella et al, A&A, 488, 2, pp. 667-674 (2008) / Linsky et al, ApJ, 455, p 670 (1995) / Dieterich et al, AJ, 147, 5, p 25 (2014).

132 / ESO Library of Stellar Spectra, A.J. Pickles, PASP 110, 863 (1998).

134 / Nearby stars, Preliminary 3rd Version (Gliese+ 1991) / Tycho-2 Catalogue, Hog et al, A&A, 355, L27 (2000).

136 / LCOGT lcogt.net/siab / Hurley et al, MNRAS, Vol. 315, Issue 3, pp. 543-569 (2000).

138 / Asiago supernova catalogue, Barbon, R., Buondí, V., Cappellaro, E., Turatto, M. 2010 VizieR Online Data Catalog, 1, 2024 / Central Bureau for Astronomical Telegrams Supernovae List (IAU, Smithsonian Astrophysical Observatory).

142 / ATNF Pulsar Catalogue, Manchester, R. N., Hobbs, G. B., Teoh, A. & Hobbs, M., The Astronomical Journal, Volume 129, Issue 4, pp. 1993-2006 (2005).

142 / Burbidge, Burbidge, Fowler & Hoyle, Rev. Mod. Phys. 29, 547 (1957).

144 / N. Capitaine et al A&A, 412, 567 (2003) / J. Lieske et al. A&A, 58, 1 (1977) / VirtualSky, LCOGT lcogt.net/virtualsky.

148 / VirtualSky, LCOGT lcogt.net/virtualsky.

150 / Fermi (NASA) / IRAS (NASA) / Planck Collaboration (ESA) / ROSAT (MPE/DLR) / Chromoscope.net.

152 / ESA / Planck Collaboration (2015).

154 / NASA / JPL-Caltech / Robert Hurt, Spitzer Science Center / NASA Fermi.

156 / McCall, MNRAS (2014) 440 (1): 405-426.

158 / Uses the 30,000 lowest redshift galaxies from the 2MASS Redshift Survey. Huchra, et al., The 2MASS Redshift Survey, ApJS.

160 / Hubble, E. P., Extragalactic nebulae, Astrophysical Journal, 64, 321-369 (1926) / Willett et al. (2013) data.galaxyzoo.org.

166 / Harrison, Cosmology: The Science of the Universe 2nd Ed, CUP (2000) / Plate XXI, Wright, An Original Theory of the Universe.

168 / ESA Gaia / Bothun, Modern Cosmological Observations and Problems, Taylor & Francis (1998).

170 / SDSS-III DR10 release (2014) www.sdss3.org/dr10/.

172 / Planck Collaboration / ESA / Structure inspired by the Millennium Simulation (Virgo Consortium).

174 / Planck Collaboration / ESA.

180 / Spacebook, LCOGT.

182-188 / PHL's Exoplanet Catalog of the Planetary Habitability Laboratory @ UPR Arecibo.

190 / SETI / NRAO / Lineweaver & Davis, Astrobiology, 2, 3, pp. 293-304 (2002) / Petigura, Howard & Marcy, PNAS, 110, 48, pp. 19273-19278 (2013) .

192 / Carl Sagan, Linda Salzman Sagan & Frank Drake / NASA.

194 / Frank Drake / Carl Sagan / Arecibo Observatory, National Astronomy and Ionosphere Center (Cornell University/NSF).

196 / PHL's Exoplanet Catalog of the Planetary Habitability Laboratory @ UPR Arecibo.

198 / gizmodo.com / New Scientist / National Geographic / Zaitsev arXiv:physics/0610031.

202 / UCLA Galactic Center Group, W.M. Keck Observatory Laser Team / NASA, ESA & John Richard (Caltech, USA) / F. Dyson, A. Eddington & C. Davidson (1920) Phil. Trans. Roy. Soc., 220, 291 / JPL Horizons, Giorgini, J.D. et al, 'JPL's On-Line Solar System Data Service', Bulletin of the American Astronomical Society 28(3), 1158 (1996) / Robert A. Brauenig www.braeunig.us/apollo/apollo11-TLI.htm.

204 / APOD created by Robert Nemiroff (MTU) & Jerry Bonnell (UMCP) / strudel.org.uk/lookUP / SIMBAD database / NASA/IPAC Extragalactic Database / SkyBoT / CBAT Supernova List / RAS of Canada Constellation List / IAU Minor Planet Center.

206 / Planetary Society / Porco et al. Science 311 1393 (2006) / Jet Propulsion Laboratory (europa.jpl.nasa.gov) / ISRO's Chandrayaan-1 & NASA / Lawrence et al., Science 339 292 (2013) / Christensen. GeoScienceWorld Elements 3 (2): 151–155 (2006).

208 / American Geophysical Union / British Antarctic Survey / Container-Transportation. com / Titanic-Titanic.com / BioNumbers.com / Cornell University.

210 / CRC Handbook of Chemistry and Physics / Kaye and Laby Online (NPL) www.kayelaby. npl.co.uk / Composition of the Human Body (Wikipedia; various sources).

212 / Earth Orientation Center of the IERS.

216 / SIMBAD database / NASA/IPAC Extragalactic Database.

218 / Dreyer, J. L. E., Memoirs of the RAS, 49, p. 1 / SIMBAD database / NASA/IPAC Extragalactic Database.

In making this book we relied on data from a large range of sources. Specific thanks go to Dr Karen Masters (University of Portsmouth), Prof Abel Méndez (University of Puerto Rico at Arecibo), Dr Lucie Green (University College London), Dr Chris Scott (University of Reading), and Dr Ryan Milligan (Queen's University Belfast/Catholic University of America/NASA Goddard Space Flight Center) for the use of data. The gas giant profiles were courtesy of Dr Leigh Fletcher (University of Oxford). Many thanks to Franziska Batten, Ben Flatman, Robert Moritz, Christine, Elisabeth Baeten, jed, Arjen van den Berg, Sophie Ward, George Williams, planet4589, Georgina McGarry, Iair Arcavi, Arman Tadjrishi, Aleks, Matthew Standing, Brooke, Andy Howell, Alex Meredith, Leon, Kyriacou, Stelios, Edward Gomez, Andi Zachai, Bernardo, Benjamin Maglio, Annie, Jake, Drew, Margaret, emu, and Cams for helping to categorise the asteroid names. The magnetic field texture was based on an image by Marc-Antoine Miville-Deschenes (IAS, Paris) and Diego Falceta-Goncalves (University of St Andrews). We made great use of various astronomical databases including: SkyBoT project (set up under the auspice of the French Ministry for National Education and CNRS); the Extrasolar Planets Encyclopedia (Jean Schneider, CNRS/LUTH - Paris Observatory); the Central Bureau for Astronomical Telegrams Supernovae List (IAU, Smithsonian Astrophysical Observatory); the Royal Astronomical Society of Canada Constellation List (Larry McNish); the Minor Planet & Comet Ephemeris Service (IAU Minor Planet Center); the NASA/IPAC Extragalactic Database operated by the JPL, Caltech, under contract with NASA; the SIMBAD database operated at CDS, Strasbourg, France.

We are also grateful for various software packages that greatly helped with the analyis and display of the data. They include: Matplotlib, a Python library for publication quality graphics (Hunter 2007); Astropy, a community-developed core Python package for Astronomy (Astropy Collaboration, 2013); HEALPix (K. Górski et al. (2005), ApJ, 622, 759) and Healpy; the WCSTools package (Jessica Mink, Smithsonian Astrophysical Observatory); PyEphem; VirtualSky/LCOGT; the VizieR catalogue access tool, CDS, Strasbourg, France; the Raphaël JS library. The collaborative nature of this project was made much smoother by using GitHub.com to work together.

The creation of this book would not have been possible without a tremendous amount of work by designer Mark McCormick - thanks Mark! Thanks also to all the designers at Founded, to Melissa Smith at Aurum Press for all her support, and to Lesley Malkin for proofreading.

Finally, but not leastly, Stuart thanks his dad, Colin, Rhona, Erica, Craig, Peter, Megan, and Ian for comments and support during the making of the book. Chris thanks his parents, Anne and Derry, and brother, Andy, for comments, and Gabi and Clara for their support and patience during the process.